STRUGGLE FOR PRESERVATION

Jon M. Cosco
Foreword by David R. Brower

Published in cooperation with the Dinosaur Nature Association

JOHNSON BOOKS: BOULDER

9 8 7 6 5 4 3 2 1

Cover design by Margaret Donharl

Cover photograph by Jeff Gnass

Map on page 10 by Martin C. Wright

Library of Congress Cataloging-in-Publication Data

Cosco, Jon M.
    Echo Park: struggle for preservation / Jon M. Cosco : published in
cooperation with the Dinosaur Nature Association.
        p.   cm.
    Includes bibliographical references and index.
    ISBN 1-55566-140-8
    1. Dinosaur National Monument (Colo. and Utah)—History.   2. Echo
Park Dam (Colo.)—History.   I. Title.
    F832.D5C67   1995
    978.8'12—dc20   95-10258
                                                          CIP

Printed in the United States by
Johnson Printing
1880 South 57th Court
Boulder, Colorado 80301

Contents

Foreword

Never having been addicted to objectivity (how much of it is a charade), I feel free to express my admiration for this extraordinary account of a conservation battle in which I was given an opportunity I could not have dreamed of—to help save Echo Park.

Occasionally, my trips from Salt Lake City to Denver have given me a chance to look down on the canyons which the Green and Yampa rivers have sculptured in Dinosaur National Monument. They are still as wild as they were in 1950, when I first learned that they were in trouble, and I feel good about their staying out of trouble. Jon Cosco is quite generous in explaining why I may be entitled to.

What I was not prepared for, when jumping at the opportunity to make these comments, was how important it would be to have this book available for those who participated in the environmental watershed of the November 1994 election, as well as those who forgot to, or simply did not, vote. They need to recognize the *déjà vu* that they have invited to reappear and confront all of us.

The first *vu* for me was at the North American Wildlife Conference in San Francisco in 1950. There I learned that something must be done to save Dinosaur National Monument to avoid a repeat of the tragedy that destroyed one of the four Yosemites in the Sierra Nevada's Hetch Hetchy Valley in Yosemite National Park. An early version of today's "wise use" movement was exercising itself—and exercising Bernard De Voto. His article in *Harper's*, "The West Against Itself," impressed me so much that I reprinted it in the *Sierra Club Bulletin*. I was glad to augment his defense of Dinosaur.

It was good to have been warned in time that something must be done. It was better to encounter evidence that more than something must be done—that it was time to drop everything else and do it. That is exactly what happened to me when I saw the home movie that Harold Bradley, who would be president of the Sierra Club when he was eighty, made of his family's float trip. They eased their way down the Yampa, through Echo Park, and on into the wild ride the Green River rewards anyone who wants to check out Split Mountain Gorge. It was an experience I wanted everyone to share, including my two young boys and me.

While the Echo Park battle raged, thousands of people did share the experience through the film Charles Eggert made for the Sierra Club, "Wilderness River Trail." Ken and Bob Brower, ages eight and six, were minor stars in it, to make the point that anyone could ride the rapids. What I didn't know at the time was that Bob would have been bounced into the Green River in Split Mountain rapids had he not been held onboard by a friend.

As I told John McPhee in *Encounters with the Archdruid,* our side never wins battles like these. We get no more than a stay of execution. The damsites are still there. The late Hugh Woodward, ardent reclamationist, told me, "Only God can make a damsite." There will always be those who don't want God's artistry lying around unexploited.

Fortunately, Dan Beard, current commissioner of the Bureau of Reclamation, is not one of those who would exploit damsites or otherwise exacerbate what Hanna Premmel of Little Rock calls the "Poverty of Materialism." But Beard is one of the first who isn't. Members of Congress who scored lowest in the ratings published by the League of Conservation Voters are now in positions of leadership in Congress. There is a challenge for all of us. For the moment, help Dan Beard hold the fort, and help those leaders realize why he must.

But beyond the moment, and for as long as this little planet retains the beauty Robinson Jeffers admired so exquisitely, let us remember that we cannot create wild rivers, we can only spare and celebrate them.

—David Brower
January 15, 1995

Introduction

In the extreme northwest corner of Colorado, a narrow fin of sandstone, some eight hundred feet high and nearly a mile long, marks the confluence of the Green and Yampa rivers. Where the two rivers meet, the Yampa releases some of the silt and sand suspended in its current. The sediment has formed a sandbar on the east side of the river, across from the sandstone monolith which rises up out of the water on the other side. On June 17, 1869, three battered wooden dories coming down the Green River beached on this sandbar. The dories were led by Major John Wesley Powell, a scientist and lecturer who had determined to navigate the lower portions of the Green and Colorado rivers and thereby fill in one of the last blank spots on the map of the United States.

Powell and his men were already more than three weeks into their expedition when they reached the mouth of the Yampa. The foreboding canyon through which they had just come was filled from wall to wall with an almost unbroken stretch of whitewater, presenting conditions so perilous that Powell and his crew lined or portaged their boats more than they rowed them. In spite of the care they took to avoid an accident, one of the dories was dashed to pieces in a violent rapid that Powell named Disaster Falls. Although the upset was only one in a series of "disasters and toils" the expedition faced, Powell hadn't lost his admiration for the river, declaring in his journal that the scenery was "beyond the power of pen to tell."[1] The rest of the oarsmen, however, were in no mood to pay the river compliments. Their clothes were wet and full of sand, and their heads addled by

View of the east side of Steamboat Rock and the junction of the Yampa (on the left) and Green rivers. August 9, 1935. *Photo by George A. Grant, Dinosaur National Monument.*

the ceaseless roar of the river. Despite the back-breaking work of lining and portaging the boats, they had not slept well because the few patches of shore in the canyon were steep and strewn with boulders.

The small sunny box canyon at the confluence of the Green and Yampa rivers was a pleasant contrast to the canyon upstream. The river was swollen and swift, but unimpeded by boulders or sudden drops. Cottonwoods, willows, and box elders lined its shore. One oarsman, impressed by the rich color of the Weber sandstone and the contrasting streaks of desert varnish, admitted to his journal that the monolith rising from the river across from the mouth of the Yampa was "the prettiest wall I have ever seen." Powell, too, was enchanted by the place and described it in his journal:

> On the east side of the river, opposite the rock, and below the Yampa, there is a little park, just large enough for a farm, already fenced with high walls of gray homogeneous sandstone. . . .
>
> Great hollow domes are seen in the eastern side of the rock, against which the Green sweeps; willows border the river; clumps of box-elder are seen; and a few cottonwoods stand at the lower end. Standing opposite the rock, our words are repeated with startling clearness, but in a soft, mellow tone, that transforms them into magical music.

Scarcely can you believe it is the echo of your own voice. In some places two or three echoes come back; in other places they repeat themselves, passing back and forth across the river between this rock and the eastern wall.[2]

Because of the manner in which the voices of his crew members reflected from it, Powell called the great sandstone wall Echo Rock. He named the scenic canyon Echo Park.

Today, Echo Park is part of Dinosaur National Monument. Here, as in other national parks and monuments, the National Park Service tries to "conserve the scenery and natural and historic objects . . . by such means as will leave them unimpaired for the enjoyment of future generations."[3] Because of these efforts, the sandbar on which the Powell expedition camped appears to the modern visitor much as it did to Powell himself. There are, of course, a few notable differences. Flaming Gorge Dam, straddling the Green River about sixty miles upstream from Echo Park, has removed some of the sediment from the river. Water released from the dam is too cold to support native fish such as the humpback chub and the Colorado River squawfish, and because the river does not fluctuate with the seasons as dramatically as it once did, spongy green tamarisk has invaded the shoreline. There have been less tangible changes, too. For instance, the sandstone promontory which Powell named Echo Rock is now called Steamboat Rock, perhaps because it reminded later visitors of a riverboat with its prow sticking high out of the water.

None of these changes drastically diminishes the aesthetic integrity of Echo Park, however, and the visitor who stands on that sandbar today can share the same sensations which so delighted Powell and his crew more than a century ago. If the season happens to be late August or early September, as it was when I first visited Echo Park, the cottonwoods and small box elders that stand along the river's edge may have begun to turn from translucent green to gold, and chances are good that the sky will be a clear, deep blue. The crowds of visitors that come through Echo Park during the summer will have thinned out, with those few remaining at the Park Service campground a few hundred yards downstream, out of earshot. Standing silently on that sandbar, it is likely that you will hear nothing except the song of the river's moving water.

The scenery by itself is impressive, but the experience will evoke deeper emotions if you happen to know that everything in sight was almost drowned beneath five hundred feet of water—you might even raise your voice and shout incredulously across the river, "They wanted to build a dam here?"

The words will echo off the rock: *Build a dam here?*

Between 1950 and 1956, that very question resonated in the corridors of the Department of Interior, in the White House, and in the halls of Congress. The question also echoed through the minds of those who were individually struggling to give meaning to the word "conservation" and through a nation that collectively was beginning to show more interest in its dwindling wilderness areas.

"Conservation" is not as easy to define as one might think. Gifford Pinchot, the forester who coined the term near the beginning of this century, believed conservation was nothing more than the "wise use" of natural resources. "The first great fact about conservation," Pinchot declared in 1910, "is that it stands for development." W. J. McGee, Pinchot's friend and colleague, defined conservation as development which resulted in "the greatest good to the greatest number for the longest time." Theodore Roosevelt picked up this utilitarian brand of conservation and made it a prominent theme in his administration. Not everyone agreed, however, that conservation should place such strong emphasis on the material values of nature. Robert Underwood Johnson, the cultured editor of *Century* magazine, believed that one of the great values of nature was its power to invigorate the human spirit. "It is much to be regretted," Johnson lamented, "that the official leaders of the conservation movement . . . have never shown a cordial, much less an aggressive interest in safeguarding our great scenery." Sierra Club founder John Muir was an even more strident advocate of the wilderness aesthetic. "Thousands of tired, nerve-shaken, over-civilized people," Muir noted, "are beginning to find out that going to the mountains is going home; that wildness is a necessity; and that mountain parks and reservations are useful not only as fountains of timber and irrigating rivers, but as fountains of life."[4]

It is by now well known that the divergent philosophies of Pinchot and Muir did not represent an idle difference of opinion; they embodied a deep and lasting paradox in the conservation movement. Their ideologies clashed for the first time over a proposal to dam Hetch Hetchy Valley in Yosemite National Park, a controversy which nearly tore apart the early conservation movement. Pinchot and other adherents to the "wise use" philosophy believed Hetch Hetchy's highest and best use was as a reservoir to supply water to San Francisco. Muir and his disciples considered it sacrilege to flood one of the most beautiful pockets of wilderness in the entire Sierra Nevada. "Dam Hetch Hetchy!" a stricken Muir wrote in response to the proposed development. "As well dam for water-tanks the people's cathedrals and churches, for no holier temple has ever been consecrated by the heart of man." However, it was much easier to describe the material benefits of water development than it was to convince policymakers of the intangible value of wilderness. Although Muir worked feverishly to arouse national sentiment for preserving the national parks, wilderness advocates did not yet have the political weight to tip the scales against the dam. In 1913, Congress authorized the project. Muir died a short time later of excessive bitterness.[5]

The Hetch Hetchy controversy made it clear that "conservation" meant different things to different people. Some "conservationists" promoted efficient use and management of natural resources, while others sought to curtail development. If the definition of conservation was uncertain, however, one thing at least was incontrovertible: conservation was—and is—fundamentally concerned with the conflicting imperatives to develop and preserve natural resources.

These conflicting imperatives have manifest themselves most dramatically in the American West, at least in part because of the vast preponderance of public lands there. The West, of course, is a loose collection of myths as much as it is a place. If one were pressed to define the region geographically, however, perhaps the most meaningful line to draw would be the one roughly corresponding to the hundredth meridian, from the Dakotas down through Texas, west of which the average rainfall is less than twenty inches a year. Twenty inches of rainfall a year, not incidentally, is the amount of precipitation needed to consistently grow crops without irrigation. It is the line, Wallace Stegner said, which separates "wet-enough" from "too-dry."[6]

Many writers, beginning with Powell, have noted the significance of western aridity. Walter Prescott Webb called the West "a semi-desert with a desert heart." Stegner agreed. "Aridity, and aridity alone," he knew, "makes the various Wests one." Of course, Stegner acknowledged that there are exceptions to the general rule of western aridity, including the western thirds of Washington and Oregon and a large part of California, where high mountain ranges wring moisture out of the air coming off the Pacific Ocean. But the vast interior of the West—eastern Oregon and Washington, southern Idaho, much of Montana and Wyoming, most of Arizona, New Mexico, Colorado, and Utah, and nearly all of Nevada—is dry, and it is this arid region that can properly be thought of as the heart of the West.[7]

Echo Park—that scenic box canyon at the confluence of the Green and Yampa rivers—happens to be very near the geographical center of the West's arid heart. It also occupies a spot near the heart of the old and ongoing debate between those who would develop the West and those who would limit its growth. In the 1950s, the United States Bureau of Reclamation proposed to build a dam at Echo Park, a dam that would be the "wheelhorse" of a comprehensive plan to develop the entire Upper Colorado River Basin. In so doing, the Bureau made Echo Park the focal point of one of the most significant conservation battles of the twentieth century.

The Bureau of Reclamation traces its roots to the National Reclamation Act of 1902. Its mission was to put rivers to work in the most efficient way possible. The Bureau's engineers considered themselves to be conservationists, but for them conservation had nothing to do with protecting the ecology of rivers or keeping them unimpaired for the enjoyment of future generations. Rivers, in their view, were to be regulated and appropriated for active use. For the Bureau's engineers, conservation simply meant controlling nature.

The Bureau's emphasis on control is hard to overstate. It is evident even in the reclamation program's earliest rhetoric. W. J. McGee, one of the earliest and most influential advocates of federal reclamation, succinctly described the impulse behind this type of conservation in 1909. "The conquest of nature," McGee declared, "which began with progressive control of the soil and its products, and passed to the minerals, is now extending to the waters on, above and beneath the surface. The conquest will not be complete until the waters are brought under complete control." Almost twenty years later, in a book entitled *Success on Irrigation Projects,* another proponent of reclamation argued the same point in even more compelling terms: "The destiny of man is to possess the whole earth; and the destiny of the earth is to be subject to man. There can be no full conquest of the earth, and no real satisfaction to humanity, if large portions of the earth remain beyond his highest control."[8]

Although in its early years the Bureau faced continuous financial difficulties and occasional political opposition, by the 1930s it had distributed enough water in the western states to establish itself firmly in the federal bureaucracy. The Depression ruined many western farmers, but it was a boon for the Bureau. Reclamation meshed well with the New Deal's efforts to put people back on their feet. The Bureau's irrigated farms and cheap hydroelectricity helped lift the country out of the Depression, and in the process helped convince the nation of the benefits of large-scale water development. Its most impressive project in that decade was Boulder—later renamed Hoover—Dam. When it was completed in 1935, Boulder Dam stood as the single most heroic engineering feat the country had ever seen. "This is an engineering victory of the first order," President Franklin Roosevelt told a crowd gathered to celebrate the dedication of the dam, "another great achievement of American resourcefulness, skill, and determination."

The accolade showered on the Bureau during the 1930s only increased during the next decade. During World War II, the Bureau's cheap hydroelectricity helped build the tanks and planes that beat back European fascism. When the war ended, the country showed its appreciation by giving it even more to do, and more money with which to do it. During these years of postwar affluence and optimism, the federal reclamation program received its highest appropriations ever—$314 million in 1950, in addition to revenue generated from the sale of public lands and electric power. The money it spent that year accounted for more than 60 percent of the entire budget for the Interior Department. A 1951 *Time* magazine article entitled "Endless Frontier" captured the nation's boundless faith in the Bureau's engineers when it declared "irrigation experts are now convinced that the rapidly growing U.S. can expand almost indefinitely within its boundaries." Nature, it seemed, could present no challenge that was beyond the ingenuity of the Bureau's engineers. With enough money and technical know-how, there was apparently no natural force which they could not tame and put to good use.[9]

The country's confidence in federal water managers was at an all-time high when the Bureau of Reclamation turned its attention to the upper reaches of the Colorado River. No other river in the United States quickened the pulse of the Bureau's engineers like the wild Colorado. Its descent of nearly fourteen thousand feet from its headwaters to its mouth at the Gulf of California is unequaled by any other river in North America. The Colorado's unregulated flow was almost lyrically erratic, fluctuating as much as a hundred fold between high and low water. It carried a tremendous amount of sediment. And along with the Green, its largest tributary, the Colorado flowed through an arid watershed, the most sparsely settled area in the continental United States. Bureau engineers considered the wild Colorado noteworthy not for what it was, however, but for what it could become. Along its banks stretched countless acres of sparsely settled land, ready to be overrun by a hardy breed of ranchers and farmers. In the minds of the Bureau's engineers, the Colorado's watershed represented nothing less than an unrealized empire—a land of promise that would sprout and blossom with the regular application of a little water.[10]

Today, as a result of the Bureau's efforts, an impressive series of dams stand on the Colorado and its tributaries—Davis, Flaming Gorge, Glen Canyon, Hoover, Imperial, and Parker, to name only a few of the largest. These dams provide a reliable source of water for irrigating dry farmland, they generate clean electric power, they store water for use in those years when rainfall is scarce, and they provide recreation for millions of people every year. But the Bureau's dams and diversion tunnels have not been kind to the river. What was at one time the wildest river in the United States is now diverted and used so extensively that it is barely more than a trickle when it reaches the Gulf of California, and that only in wet years; in dry years, the Colorado actually disappears in its own sandy delta, miles from the sea.

As author Marc Reisner has noted, the present condition of the river evokes strong emotions for many people, emotions that range from pride to despair and outrage. "To some conservationists," Reisner remarked, "the Colorado River is the preeminent symbol of everything mankind has done wrong. . . . To its preeminent impounder, the U.S. Bureau of Reclamation, it is the perfection of an ideal."[11] Ultimately, the emotions one feels depend upon the system of values with which one evaluates the objective facts. A romantic longing to preserve the river's scenic wildness conflicts with the practical necessity of utilizing its water wisely and efficiently.

These values collided dramatically when the Bureau of Reclamation proposed to build a high dam at Echo Park. With its sheer sandstone walls, Echo Park was an excellent site for a dam. E. O. Larson, an engineer for the Bureau of Reclamation, once referred with a certain awe to "the remarkable storage vessel at Echo Park." But Echo Park was also a wonderfully scenic spot, and part of the national park system. When David Brower, who led opposition to the dam, first floated through the canyons of Dinosaur National Monument in the summer of 1953, he reported with an awe of his own, "I have never had a scenic experience equal to that one."[12] The spiritual and aesthetic values which shaped Brower's opinion of the river and its canyons were clearly different from the technological and utilitarian values which informed Larson's opinion of the same. Similarly, the values which led some groups to oppose the Echo Park dam conflicted with the values that led other groups to support it.

When the issue of debate is a single dam, there is little room for compromise—the dam is either built, or it isn't. Benton MacKaye, one of the original cofounders of the Wilderness Society, suggested this in a memorable manner in the spring of 1954, when he learned that the Echo Park dam would create a reservoir averaging two hundred feet deep, in canyons that exceeded three thousand feet. "Does this mean that the 'scenic values' of the canyon walls would be impaired by only 6.67 percent?" MacKaye wondered. "Or might it be 100 percent? Is this innocent appearing reservoir pleading, in accordance with the parable, that it is only 'a wee bit pregnant'?" To MacKaye, and other opponents of the dam, such a claim was ridiculous. "In this case, the demand is for one thing or some other," declared MacKaye, "for all or nothing. A canyon, like a cake, you cannot have, and bite it too."[13]

In politics, however, opportunities for compromise can sometimes emerge unexpectedly. Despite Benton MacKaye's insistence that the Echo Park controversy was "for all or nothing," after five years of political stalemate conservationists found themselves staring at a deal that would save Echo Park. The Bureau of Reclamation would drop its plan to build a dam in Dinosaur National Monument if, in return, conservationists agreed to support its comprehensive plan to develop water resources in the Upper Colorado Basin. By accepting this compromise, conservationists would save Echo Park and preserve the principle that the national parks, once designated, were inviolable.

Never before had wilderness advocates enjoyed the political muscle to prevent such large-scale development of a natural resource—especially when the resource was water and the setting was the arid West. To be sure, there had been conservationists who could write eloquent pleas for nature and influence public opinion; there had been conservationists who enjoyed special influence with Washington officials; and there had even been policymakers who considered themselves "conservationists" of one sort or another. But throughout the first half of the twentieth century, conservation organizations lacked the kind of organized constituency that could consistently sway the policy decisions of congressmen and cabinet secretaries. The Hetch Hetchy battle had demonstrated that point conclusively.

But since the damming of Hetch Hetchy, the conservation movement had been quietly transforming itself, like a caterpillar inside its cocoon. The Echo Park controversy marked the moment at which it emerged into the light of national politics. In their fight to save Dinosaur National Monument, conservationists demonstrated an unprecedented solidarity, an adeptness at lobbying, and a heightened awareness of the economic and technical arguments that had been used for so long by the opposition. By coming together, they gained a political voice. For that reason alone, the Echo Park controversy is historically significant.

But the story of Echo Park is as at least as important for what it suggests about the environmental movement today. In the fight to save Echo Park, conservationists learned that a political voice comes with a responsibility to use its power appropriately. Those who wrestled with that responsibility in the 1950s faced a set of issues that has hardly changed in the four decades since. Today, perhaps more than ever before, environmentalists must struggle with the practical and ethical questions that arise in the political arena. They must decide when it is appropriate to negotiate and when compromise is unacceptable.

The Echo Park dam controversy presented those questions in particularly compelling terms. By illuminating the answers that conservationists put forth in the 1950s, we may better understand the context in which we try to resolve our own environmental dilemmas. John McPhee called the battle to keep dams out of Dinosaur National Monument "a milestone in conservation history."[14] I hope to show that his casual remark contains at least a little truth. If, with the benefit of hindsight, we can claim that the modern environmental movement had some clearly recognizable beginning, we might start to look for it in a scenic canyon called Echo Park.

CHAPTER ONE

Growth of a Dinosaur

D inosaur National Monument, straddling the border of Utah and
Colorado, has long been and still remains one of the least
known units in the national park system. Its obscurity is partly the
result of its remoteness. To get to Dinosaur from the closest urban centers requires a drive on U.S. 40 of about four hours from Salt Lake City
and more than six hours from Denver. The drive from Denver is a
quiet one. After climbing over the Continental Divide at Berthoud
Pass, the road winds across Rabbit Ears toward Steamboat Springs.
Steamboat is a renowned ski resort, and its lodges and specialty shops
lend the town a boomtime atmosphere, even in the summer. But after
Steamboat the road traverses a high plateau, cutting through hundreds
of thousands of acres of dry, rolling rangeland, full of sagebrush and
graced only by the occasional cow. Perched precariously in the middle of this high plateau, about midway between Steamboat Springs
and Dinosaur, is the small city of Craig. Unlike Steamboat Springs,
there is nothing glamorous about Craig. Ranching dominated the economic activity there until the 1930s, when the range deteriorated and
many residents found it more profitable to switch to coal mining and
associated industries.

The route from Salt Lake City winds through the Wasatch Mountains
and after passing close to ski resorts such as Park City, the U.S. 40
enters the Uinta National Forest. The road then roughly parallels the
southern front of the Uinta Mountains, one of the few east-west trend-
ing ranges in North America. The flat, rocky expanse at the foot of
the mountains, known as the Uinta Basin, is so dry and desolate that

no one, not even the hardy Mormon pioneers, dared settle it. Mormons sent in 1861 to explore the area reported to the church fathers that it was "measurably valueless . . . except to hold the world together." The federal government didn't want the land either, so that same year it declared the Uinta Basin a permanent reservation for the Ouray and Ute Indians. In 1906, at the urging of ranchers and railroad companies, the federal government changed its mind and opened the area to white homesteaders. Some of the towns established by these homesteaders still exist along U.S. 40, including Duchesne, Roosevelt, and Vernal.

Vernal, Utah, is still the last place to buy a new tire or a sit-down dinner before reaching Dinosaur National Monument. In its early days, Vernal was as lively a cow town as the West has seen. As late as the 1940s, when less than four thousand people lived in Vernal, men walking down main street were likely to be wearing broad-brimmed Stetsons, high-heeled boots, and jangling spurs. But like Craig and other nearby towns, Vernal gradually turned to the extraction and refining of fossil fuels. About 7,300 people now live in Vernal. The town boasts many small businesses, but no large ones, and on the few irrigated farms scattered around Vernal, the predominant crop is alfalfa, grown as feed for livestock.

Dinosaur's obscurity is primarily a function of its remote setting; the monument is, as residents of Craig like to say about their town, "a mountain pass away from everything." But the name Dinosaur National Monument is something of a misnomer, and this has undoubtedly contributed to its obscurity.

To be sure, President Woodrow Wilson created the original eighty-acre monument in 1915 to protect a quarry of fossilized dinosaur bones discovered by paleontologist Earl Douglass. The museum which the National Park Service eventually constructed around the quarry's sandstone outcropping draws hundreds of thousands of visitors every year. But Dinosaur is much more than fossilized bones. It encompasses more than 209,000 acres of undeveloped land, including spectacular canyons carved by the Green and Yampa rivers. Unlike the boundaries of the original eighty-acre preserve, which were arbitrarily laid out in a square around the Douglass quarry, Dinosaur's present boundaries show a close relation to the river canyons which form the heart of the monument. Although its classification as a national monument implies that Dinosaur is only a junior member in the national park system, it is in fact larger than many of the better known national parks, including nearby Arches, Bryce Canyon, and Zion National Parks.[1] To the people who know Dinosaur, there is no question that its primary value is that of all national parks: a place where people can recreate or enjoy a bit of solitude amidst spectacular, untrammeled natural scenery.

2

Dinosaur grew to its current size in 1938, when Secretary of Interior Harold L. Ickes persuaded Franklin Roosevelt to expand its boundaries through executive order. As the principle steward of the public domain, Ickes was eager to establish a more prominent federal role in the management of its timber, range, and water resources. Fortunately, the Roosevelt administration was more than willing to reconsider the federal government's traditional laissez-faire approach to the public lands. Within two years of becoming president, Roosevelt had signed the Taylor Grazing Act, effectively reversing a seventy-year-old policy of federal giveaway of the public domain. Ickes, always on the lookout to increase the prestige of his own department, saw in this action an opportunity to transfer some of the more valuable portions of the public domain to various bureaus and agencies in the Interior Department.[2]

Foremost among the agencies Ickes wanted to expand was the National Park Service. Unlike most of his predecessors at Interior, Ickes was a strong advocate of wilderness preservation, an idea that at the time was quite new and, for many people, quite unacceptable in a country that still had vast, sparsely settled areas in its arid interior. In 1934, Secretary Ickes gathered all of the national park superintendents in Washington to explain his "general attitude on what our national parks ought to be":

> I do not want any Coney Island. I want as much wilderness, as much nature maintained and preserved as possible. . . . I recognize that a great many people, an increasing number every year, take their nature from the automobile. I am more or less in that class now on account of age and obesity. But I think the parks ought to be for people who love to camp and love to hike and who like to ride horseback and wander about and have . . . a renewed communion with Nature. . . . I am afraid we are getting gradually alienated from that ideal. We are becoming a little highbrow; we have too many roads. We lie awake nights wondering whether we are giving the customers all of the entertainment and all of the modern improvements that they think they ought to have. But let's keep away from that, because if we once get started, there will be no end.[3]

Ickes's statement not only displayed a remarkable sensitivity to the state of the park system, but also foreshadowed a conflict over the purpose and meaning of the national parks. That conflict was already simmering in 1934, but it would boil over in the postwar affluence of the 1950s.

Fortunately for Ickes, President Roosevelt shared his love of nature and his interest in the park system. In June 1933, Roosevelt signed an executive order that more than doubled the size of the national park system by transferring to it sixty-four national monuments, military parks, battlefield sites, cemeteries, and memorials from other agencies. The Park Service also took control of the land surrounding Lake Mead, the reservoir behind Boulder Dam, to administer as a national recreation area. Together Ickes and Roosevelt prodded Congress to establish several new national parks, including Great Smoky Mountains and Everglades in 1934 and Big Bend in 1935. Other additions, such as Olympic and Kings Canyon, would follow in 1938 and 1940, respectively. When Ickes saw little hope of congressional action, he turned to Roosevelt to expand the park system through executive order under authority of the Antiquities Act of 1906. Roosevelt obliged, creating in the three years between 1935 and 1938 dozens of new national monuments.[4]

Ickes and Roosevelt focused their attention on the arid portions of the public domain, particularly the desert southwest and the Colorado River Basin. Ickes must have had an affinity for the slickrock canyons along the Colorado, because he wanted to include them all in the national park system. His proposal for an Escalante National Monument would have set aside an almost continuous series of canyons stretching from the confluence of the Green and Colorado rivers to a point just north of the Utah-Arizona border. At 4.4 million acres, Escalante would have been by far the largest unit in the national park system. Although the proposal met fierce resistance from grazing and irrigation interests in Utah and eventually had to be dropped, Ickes did not neglect opportunities to add less ambitious parcels to the park system.

One of the areas that caught his eye was the eighty-acre preserve known as Dinosaur National Monument. Ickes believed a much larger area around the fossil quarry deserved protection for its scenic and historical value. Topographically, the area Ickes had in mind is defined by the deep canyons which the Green and Yampa rivers have cut through an eastward extension of the Uinta Mountains. The benchland surrounding these canyons includes a wide array of geological formations, from wide, flat-topped mesas to sharp, hogback ridges, ranging in age from the Uinta Mountain quartzite of the Precambrian period to the Browns Park sandstone of the Pliocene. The most interesting geology, however, is found in the canyons themselves, where the Green and Yampa rivers exposed dramatic anticlines and layers of sedimentary rock. These features provided geologists with a rare opportunity to understand the forces which shape the landscape. In fact, many of the geologists responsible for advancing their science in the latter half of the nineteenth century worked in the Uinta Mountains in and around Dinosaur National Monument. They left their names—King, Hayden, and Marsh—on some of the highest peaks in that range.[5]

The human history of the area was also of special interest. Indigenous people had occupied the canyons of the Green and Yampa rivers for many centuries. The monument stands near the confluence of three major indigenous cultural areas—the Great Basin, the Plains, and the Southwest. Archeologists have since classified the indigenous culture as the Fremont, the northernmost of the Anasazi subcultures. Archeologists still debate the roots of the Fremont culture and when it declined. Whatever mysteries remain about the Fremont, however, cannot be attributed to a lack of evidence. They left a wealth of archeological sites in and around Dinosaur's canyons, as well as copious pictographs and petroglyphs on the cliff faces.[6]

The first Europeans known to pass through the area were Spanish missionaries who crossed the Green River south of the Uintas in 1776, near the present-day town of Jensen, Utah. By the middle of the nineteenth century, a few more Europeans had undoubtedly passed through the area, in search of furs or an overland route to California.[7] But the canyons of the Green and Colorado rivers were essentially unknown until 1869, when John Wesley Powell led an expedition down the Green River to its junction with the Colorado, and eventually through the Grand Canyon. With no government support other than the right to draw rations from western army posts, Powell completed the journey of over a thousand river miles, naming many of the area's prominent natural features on the way. He made a second voyage in 1871 and then set to work systematically surveying and mapping the great slickrock canyon country, work that would culminate in his 1878 *Report on the Lands of the Arid Regions of the United States.*[8]

Powell's *Report* explained that the scarcity of water would limit settlement in Utah and the rest of the arid lands. The *Report* drew much attention—and then derision—from specialists in irrigation and water management, who were still too optimistic to accept any limits to westward expansion. Although Powell's ideas would not receive the acclaim they deserved for many years to come, his account of the exploration of the Colorado enjoyed more widespread, popular recognition. In 1874 and 1875, *Scribner's Monthly* ran a lavishly illustrated series of articles about the 1869 expedition. Audiences in eastern cities and towns, hungry for accounts of the western frontier, devoured Powell's articles and asked for more. By 1915, Powell's account of the expedition had been published twice in book form, and a large audience had become acquainted with his adventures in the Green River canyons that Ickes wanted to add to Dinosaur National Monument.[9]

Although the enlarged monument would include quite a bit of elevated benchland, no one doubted that the most important sections of the proposed monument were the canyons, particularly the areas known as Split Mountain and Echo Park. A 1935 report prepared by Herbert Evison, a landscape architect for the National Park Service, said of Split Mountain, "Geologically and scenically, this is probably the most extraordinary feature in the entire monument." Evison's description of Echo Park, locally known as Pats Hole, was equally enthusiastic: "The area at the junction of the two rivers, in sheer loveliness, exceeds anything else in the whole area, the fresh green of the flats of Pats Hole contrasting so sharply and beautifully with the rocky walls that lie above them."[10]

These canyons were accessible only by a long hike through dry, inhospitable country or by the rivers that had carved them. Since the river was not easily navigated, few people had visited the canyons after Powell. In fact, the sheer-walled canyons proved to be an excellent place to hide rustled cattle, and for many years outlaws of all descriptions took up residence out of reach of the law in an area known as Browns Park, just to the north of the proposed monument. Because the canyons were so inaccessible, most residents of Utah and Colorado believed that their only significant value was as potential water or power development sites.

By the early 1930s, the Utah Power and Light Company had already submitted an application for a power site at Echo Park to the Federal Power Commission. When Ickes announced his proposal to add the river canyons to Dinosaur National Monument, however, the power company voluntarily withdrew its application. To ensure that demand for future power development would not interfere with his plan to reclassify the area as a national monument, Ickes attempted to persuade the Federal Power Commission that the power resources in the area were insignificant. As evidence, he pointed to the utility's voluntary withdrawal. Despite Ickes's testimony, however, the commission refused to relinquish the right to develop power resources in the area. Its ruling stated that "regardless of the disposition which may be made of the Utah Power and Light Company's application . . . the Commission believes that the public interest in this major power resource is too great to permit its impairment by voluntary relinquishment of two units in the center of the scheme." With this disclaimer, the commission agreed to go along with the enlargement, provided the executive proclamation contain "a specific provision that power development under the provisions of the Federal Water Power Act will be permitted."[11]

Members of the Federal Power Commission were not the only people Ickes had to win over, however; local residents also expressed concern about permanently "locking away" resources that might be needed for economic growth at some point in the future. Local cattlemen supported the expansion, but only because they mistakenly believed the land reclassification would enhance grazing in the area by removing it from the jurisdiction of the new Taylor Grazing Act. To safeguard the state's access to local resources, Utah Governor Henry Blood and Senator William King both requested that specific recognition of mineral and reservoir rights be put into the proclamation.[12] Ickes was reluctant to make any specific reservations, but he agreed that future development should be determined by Congress if and when it was economically feasible. A Park Service official named David Madsen probably confused the issue when he assured residents of nearby Vernal, Utah, and Craig, Colorado, that the enlargement of the monument would in no way affect the future development of water projects in the area. Although Ickes had explicitly prohibited any Park Service personnel from making such promises, Madsen later claimed that all of his statements had been authorized by Park Service superiors. Whether or not Madsen was speaking officially for the Park Service, the local audience certainly believed he was. Several local newspapers, including the *Salt Lake Tribune,* reported that the Park Service would permit grazing and power development in the new monument. Content that their economic interests were not being compromised, the citizens of Vernal approved the enlargement of the monument by petition.[13]

President Roosevelt issued the proclamation expanding Dinosaur National Monument on July 14, 1938. In response to local demands, the proclamation included a provision stating, "This reservation shall not affect the operation of the Federal Water Power Act of June 10, 1920, as amended, and the administration of the monument shall be subject to the Reclamation Withdrawal of October 17, 1904, for the Browns Park Reservoir Site."[14] The wording of the proclamation later proved to be somewhat ambiguous. Browns Park, near the northern end of the expanded monument, was a relatively open area, a poor site for the construction of a dam when compared to some of the sheer-walled canyons further downstream, and it is unclear why this area was singled out by name. No specific mention was made in the proclamation regarding more desirable sites at Echo Park or Split Mountain, although it was understood that Congress could authorize a project in either of those areas if it so desired. In addition, there seems to have been some confusion about the terms of the Federal Power Act. When it was initially passed in June 1920, the act allowed the construction of hydroelectric projects in the national parks and monuments system. One of the first acts of the next session of Congress, however, had been to amend the act to explicitly prohibit such projects in the park system. In 1935, another amendment was added to strengthen and clarify this intention. If the proclamation's reference to the Federal Power Act was intended to protect water development rights in Dinosaur National Monument, it was misguided and misleading.[15]

The Park Service, in any case, showed little concern about the issue of future water development. Perhaps overly elated with all of the recent additions to the park system, it seemed willing to make whatever concessions were necessary to appease local concerns. The Browns Park reservation, for instance, conflicted with a Park Service report which stated emphatically that "unless the power rights in this area are restricted, the National Park Service should not consider this area as a national monument or national park."[16] David Madsen's promises to local residents were probably based at least partly on the assumption that there was no need and no desire to develop a power site in the area. Conrad Wirth, at the time an assistant director of the Park Service, neatly captured the nonchalant mood of the Park Service. There was no need to make a decision about reclamation or power use, Wirth believed, since these would "give relatively little difficulty" in the future. Local demand for irrigation or power was not great enough "to force the construction of reservoirs within the proposed monument, at least not for a great number of years."[17]

Before long, Wirth would have to ruefully acknowledge how faulty his judgment on this issue had been.

CHAPTER TWO

"Here Comes a Boom"

O n January 2, 1939, a truck loaded with surveying equipment pulled off U.S. 40 near the small town of Jensen, Utah, and headed up a rutted road toward the headquarters of Dinosaur National Monument. The engineers and geologists in the truck were on their way to a canyon known as Split Mountain, the southernmost in a series of canyons through which the Green River charged before flowing peacefully out into the flat expanse of the Uinta Basin. They would have to unload their truck at the Park Service headquarters and hike about a mile and a half to the mouth of the canyon, because, as yet, there was no road to the site where they wanted to build their dam.[1]

Any resident of Jensen who might have seen the truck rumble by and noticed the emblem emblazoned on its side panel—a dam, a reservoir, and an irrigated field all inside a mighty drop of water—probably would not have recognized it as the insignia of the United States Bureau of Reclamation. Although the Bureau had assisted in some major irrigation projects in Utah, up until this time it had concentrated its most impressive efforts in California's Imperial Valley and on the lower reaches of the Colorado River. The bulk of irrigated farmland in Utah was the result of small farmers working under the paternalistic, watchful eye of the Mormon Church rather than the intervention of the United States government.[2]

Indeed, most residents of Utah were probably not aware that the Bureau of Reclamation had fixed its eyes on the water of the Green River. As the federal agency charged with delivering water to western farmers, the Bureau had always had a general interest in the Green.

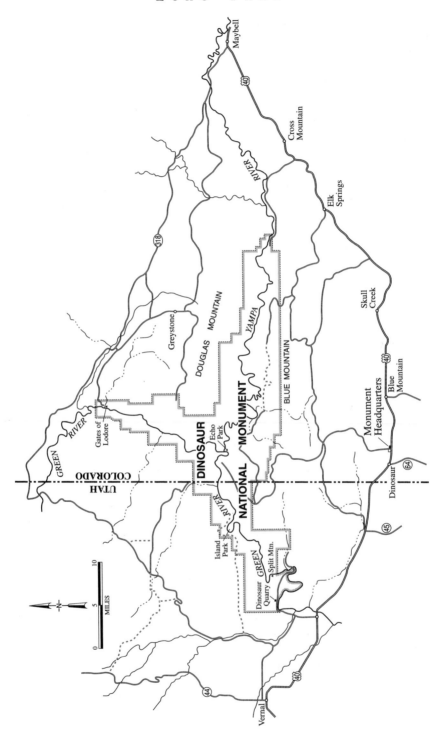

That interest had piqued dramatically, however, with the adoption of the Colorado River Compact in 1922. Assuming an annual flow somewhere close to 17.5-million-acre-feet a year, the compact had arbitrarily divided the Colorado River basin at Lee's Ferry, Arizona. Under the terms of the compact, the Upper and Lower basins were each guaranteed 7.5-million-acre-feet of water a year. The compact said nothing about how the states of each basin were to divide the allocated water among themselves, leaving the states to argue about that. In 1928, in spite of the objections of the state of Arizona, Congress approved the compact. That same year, Congress authorized the construction of Boulder Dam and an "All-American" canal to divert up to 4.4-million-acre-feet of the Lower Basin's water to California. With these two projects, the Bureau of Reclamation began to develop the Lower Colorado River Basin.[3]

A comprehensive plan for developing the Upper Basin would have to wait until the states of Wyoming, Utah, Colorado, and New Mexico could settle upon a formula for the fair allocation of the available water. However, rather than sit idly by while the states quarreled, the Bureau of Reclamation had begun in the mid-thirties to investigate potential dam sites throughout the Upper Basin. By 1935, the Bureau and the U.S. Geological Survey had focused considerable attention on the Colorado's longest tributary, the Green. Commissioner of Reclamation John Page probably knew that his boss, Secretary of Interior Harold Ickes, was planning to add much of the Green River to Dinosaur National Monument. Page undoubtedly was also aware that under the terms of the Federal Power Act such a reclassification would prevent all water projects within the monument's boundaries. That knowledge, however, did not prevent the Bureau from identifying several potential dam sites on the Green. Split Mountain, the most desirable site, remained on the Bureau's list even after the 1938 proclamation expanding Dinosaur National Monument made it part of the national park system.

The Bureau of Reclamation technicians who rumbled up to the monument's headquarters in January 1939 seemed unconcerned that Split Mountain was part of the newly enlarged monument. If he had his own concerns about the arrival of Bureau technicians, Dinosaur's Assistant Superintendent John McLaughlin did not want to sound too indignant when he wrote a few days later to the Bureau's Denver office asking for an explanation. McLaughlin, too, was undoubtedly aware of the 1921 and 1935 amendments to the Federal Water Power Act which made national parks and monuments off-limits to water projects, and he was irked that the Bureau technicians were planning to conduct their studies as if they had no knowledge of these restrictions. In his letter to the Bureau's Denver office, McLaughlin politely acknowledged that "certain reservations were made in the executive order which would

First Dam Site at Split Mt. Canyon on Green River near Jensen Uinta Co. Utah.

For many years prior to the Echo Park dam controversy, the Bureau of Reclamation investigated the geologic situation at potential dam sites in order to make their recommendations to Congress. The Bureau dug test pits in Split Mountain to determine if this site was appropriate for a dam. *Dinosaur National Monument.*

permit such exploration as the Bureau of Reclamation apparently contemplates," but he made his concern clear by requesting that "we should sincerely appreciate it if you would inform us as to the nature and extent of the work you have in mind within the monument." When the Bureau's Denver office replied that they had long considered Split Mountain to be one of the most attractive power sites on the Green River, that preliminary geological studies were close to completion, and that further investigations were being planned for areas upstream from Split Mountain, McLaughlin's concern rose another degree. He was especially disturbed by the Bureau's request that the Park Service identify those locations of outstanding archeological or scientific value so that "any plan of development finally adopted might avoid [their] injury." Not wanting to pursue the matter on his own, McLaughlin forwarded this news to the director of the National Park Service in Washington, D.C., and pointed out that the outstanding features of the monument were the canyons themselves in which the Bureau's "preliminary investigations" were being carried out.[4]

If McLaughlin expected precedent to guide Director Arno Cammerer's response to the Bureau's studies, then he probably believed his superior would take a strong stand in defense of the monument. Since its creation in 1916, water development projects had been anathema to the National Park Service. In fact, the legislation which created the National Park Service was in large measure a response to the 1913 decision to dam Hetch Hetchy Valley in Yosemite National Park. Stephen Mather, the Park Service's first director, had successfully used the legal authority in this enabling act to ward off a number of water development projects in the national parks in the 1920s. Most notable, perhaps, was Mather's fight to abrogate water-rights applications by the city of Los Angeles for streams in Yosemite and Kings Canyon national parks. With the help of his protégé, Horace Albright, Mather also thwarted plans of Idaho irrigationists to draw water from Yellowstone National Park. In Mather's mind, it was imperative to prevent any and all water development projects in the park system. "Once a small dam is authorized for irrigation or other purposes, other dams will follow," he predicted. "Once [we] start the national parks toward national forest status, it will be logically impossible to stop short. One misstep is fatal."[5]

The Park Service had not successfully opposed all water development schemes which threatened the integrity of the park system. Projects planned by private utility companies or municipalities were one thing; projects planned and promoted by an agency of the federal government were quite another. Only four years earlier, in 1935, the Bureau had proposed a water diversion project in Colorado's Rocky Mountain National Park. The Big Thompson project, as it was known, would divert headwaters of the Colorado River through a tunnel built under the park to the Big Thompson River on the other side of the Continental Divide. Like Dinosaur, Rocky Mountain had been created with specific provisions concerning future water development. Albright, having now succeeded Mather as director of the Park Service, was afraid the diversion tunnel would set a bad precedent of development in the park system and appealed to Secretary Ickes to oppose the project. Ickes agreed that he too was "afraid of establishing a bad precedent," but pointed out that "the precedent really was set when the Government accepted the park with conditions attached." Ickes eventually approved the project, telling a group of conservationists, "I have to follow the law. I tell you very frankly that between the Bureau of Reclamation and the Park Service, I am for the Parks, but I am sworn to uphold the law. . . . I wish the baby had not been laid on my doorstep, but it is there."[6]

Undoubtedly, Assistant Director Arthur Demaray had much of this history in mind when he replied to McLaughlin's memorandum describing the Bureau of Reclamation's investigations at Dinosaur. The situation at Dinosaur must have seemed to Demaray dangerously similar to the scenario surrounding the Big Thompson project. After all, the executive proclamation enlarging the monument *did* contain a provision allowing some form of power development within Dinosaur, similar to the provisions that had led to the authorization of the Big Thompson project. In light of this fact, Demaray did not seem sure how vigorously the Park Service could object to Bureau's surveys at Dinosaur. "There seems to be no question of the authority of the Bureau of Reclamation to make the present studies within Dinosaur

National Park Service
Assistant Director
Arthur Demaray.
*National Park Service,
Historic Photograph
Collection.*

National Monument and to report their findings to Congress," he wrote to McLaughlin in February. Not wanting to concede any more than was necessary, Demaray added, "It is doubtful, however, that there is any authority for construction of any dams in the monument without further authorization from Congress." Resigned to the fact that nothing could be done to halt the Bureau's studies for the time being, but confident that no construction would take place without the consent of the Congress, McLaughlin and Dinosaur Superintendent David Canfield waited as patiently as they could. Canfield, for one, did "not see that anything would be gained" by having the Park Service oppose surveys in the monument, at least "until definite information has been made available."[7]

As the Bureau continued its initial surveys at Split Mountain, it soon became apparent that there was very little the Park Service could do to *proactively* impede the Bureau's investigations. The Bureau could therefore control the pace at which any ensuing controversy would unfold. This advantage was heightened by the fact that the Park Service was dependent on the Bureau for all technical information concerning the feasibility of potential dam sites and updates on the progress of the investigative work. Bureau officials were either slow to generate this information or reluctant to pass it on to the Park Service. By the spring of 1940, Park Service officials privately began to show some exasperation at the Bureau's unwillingness to keep up cooperative correspondence. On March 29, a Bureau official explained apologetically that one Park Service request for information had never been received, but even at this late date the official stated he "could make no definitive statement regarding construction," adding only that he believed the Bureau would begin work on one or more power dams in "a matter of a relatively few years."[8]

As it became clear that the Bureau of Reclamation was tightening its grasp on the Dinosaur dam sites, new Park Service Director Newton Drury began to take an active interest in the situation. Drury had formerly been director of the Save-the-Redwoods League in California and was widely known to be a staunch advocate of wilderness.[9] He believed his primary task as director was to prevent roads and other commercial developments from infringing upon the parks. He did not seek large appropriations from Congress because he saw no reason to fatten the Park Service's budget. "We have no money," Drury was fond of saying of his agency, "therefore we can do no harm."[10]

However, Drury was savvy enough to realize that a small budget translated directly into a lack of political clout, and the Park Service director knew his agency was not likely to win a fight with the Bureau of Reclamation. Only a few years earlier, in 1936, the Bureau had convinced Congress to give it a yearly appropriation of $16 million on top of the revenues it received from the sale of public lands and hydroelectric power. A year later, the Bureau was put in charge of California's Central Valley Project, which at the time was the most expensive and ambitious irrigation project ever undertaken. By 1939, the Bureau's annual appropriations had doubled to $32 million, and the next year they doubled *again* to $64 million. The Bureau also had strong regional constituencies to support its projects while the Park Service, in contrast, had a very diffuse, national constituency that would be extremely difficult to mobilize. In any case, the complex technical and economic nature of the Bureau's projects effectively cloaked them from intense public scrutiny.[11] With the country poised on the brink of a world war, Drury knew that the secretary of Interior would clearly place priority on the development of the nation's resources. When the United States entered the war a few months later, the balance of power in the Interior Department would be confirmed as nonessential government agencies, including the Park Service, were moved to Chicago to make room for defense agencies. The Bureau of Reclamation stayed in Washington and began to justify studies in Dinosaur and other units of the park system as necessary to ensure that the country would have sufficient power production throughout the war.[12]

A dam in Split Mountain was first proposed but then dropped in favor of the Echo Park dam. The original proposal for the Split Mountain dam would have inundated a unique scene where the Green River cleaves the Split Mountain anticline, revealing colorful Mississippian and Pennsylvanian formations 290 to 360 million years old. *Photo by Jack Boucher, Dinosaur National Monument.*

Sensing that the Bureau's grasp on the Dinosaur dam sites would be difficult to loosen, Drury began to look for a compromise in which he could best protect the Park Service's long-range interests. If there was no way to stop the dams from being built within the boundaries of the monument, Drury believed, the only way to avoid a dangerous precedent of violating the national park system was to simply redraw the boundaries. As early as June 1940, a regional officer of the Park Service suggested that the Park Service would give "serious consideration" to alterations in Dinosaur's boundaries "where it is logical to do so."[13] By 1941, Drury similarly felt that the Park Service must respond reasonably to the local desire for water development to avoid a direct confrontation with the Bureau of Reclamation. Rather than risk losing Dinosaur entirely, Drury thought it the wiser course to cooperate with the Bureau and hope to influence its final plan.

On November 4, 1941, Drury and Reclamation Commissioner John Page signed a secret "memorandum of understanding" which essentially stated that the Park Service would not interfere with water projects in Dinosaur National Monument or in Grand Canyon National Monument. Drury "agreed in principle" to a change in the designation of the latter from a national monument to a national recreation area and said that "it seems not improbable that a policy similar to that already agreed to in principle . . . could be applied" to Dinosaur as well. Although legislation would be required for such a change, Drury assured Page that the Park Service did "not believe such legislation would be difficult to secure."[14]

In March of the following year, Drury and Page were once again in communication about the status of Dinosaur. This time Page stated that his Bureau was preparing a report on the Green River and wanted reassurance that the Park Service would cooperate in removing all obstacles to a water project in Dinosaur. Drury again conceded that "the time may come when it will be necessary to seek legislation to change the status of Grand Canyon and Dinosaur National Monuments," but he tried to keep the issue unresolved by stating that "it would be better not to emphasize the land change statuses at this time." Although he undoubtedly felt his only course of action was to cooperate as much as possible and stall for time, the "memorandum of understanding" would cause considerable embarrassment for the director at a later date. Drury may have foreseen some of the criticism that would be directed at him when he asked Page to keep the memorandum confidential, stating, "Anything placed in your report regarding this [agreement] would almost certainly result in immediate confusion of motives and objectives in the minds of many conservation groups."[15] In fact, conservation groups would be even more confused about Drury's "motives and objectives" when the agreement was revealed to the secretary of Interior eight-and-a-half years after Drury signed it.

In the *Federal Register* of July 13, 1943, the Bureau of Reclamation officially published power withdrawals at the Echo Park and Split Mountain sites, both well within the confines of Dinosaur National Monument.[16] In violation of a standard procedure for intra-agency communication in the Department of Interior, the Bureau failed to notify the Park Service of these withdrawals. Presumably the Park Service was aware of published notices in the *Federal Register*, but the power withdrawals somehow escaped the attention of Park Service officials until December.[17] When the withdrawals were discovered, the lack of notification seemed to confirm earlier complaints that the Bureau was withholding information and irresponsibly promoting its own projects. Yet the Park Service responded nonchalantly both to the procedural error and to the threat of development. Drury probably did not press the matter because he had been in informal contact with Page and had already agreed not to oppose the projects, although he did write to the commissioner to express concern that the withdrawals had been published before removing the affected area from the monument.[18] Regional officials who had no knowledge of Drury's secret memorandum and little understanding of the Bureau of Reclamation's political clout did not press the issue, perhaps because they were confident the monument would be protected by "changing economic conditions and the growth of strong public opinion favoring the preservation of wilderness."[19]

Indeed, the sentiment for preserving wilderness had grown considerably in the last two decades, both in government agencies and in the general public. In 1919, landscape architect Arthur Carhart proposed the seemingly radical idea that the wisest use of some U.S. Forest Service wilderness areas was to leave them undeveloped. Seven years later, at the urging of Aldo Leopold, the Forest Service designated its first official primitive area, and by 1939, the wilderness idea was institutionalized by the Forest Service's U-1 and U-2 regulations, which restricted roads, settlement, and economic development on some fourteen million acres of public land. During the same time period, popular interest in wilderness preservation and outdoor recreation was evident in the establishment of several private organizations, including the Izaak Walton League, founded in 1922, and the Wilderness Society and the National Wildlife Federation, both founded in 1935.[20] Scientists, too, began to urge wilderness preservation for ecological as well as aesthetic and recreational considerations. "All wilderness areas . . . have a large value to land science," stated Aldo Leopold, president of the Ecological Society of America, in 1941. "Recreation is not their only, or even their principal, utility."[21]

Yet however much sentiment for wilderness preservation may have grown in the preceding decades, that sentiment had not taken a very strong hold in Utah in the 1940s. The Park Service's response to the Echo Park and Split Mountain withdrawals probably would not have been so casual had officials considered that the issue was still almost entirely confined to the Upper Basin states. For the people who lived in that arid climate, water to irrigate dry fields was more important than preserving natural scenery. Salt Lake City's *Deseret News* would succinctly describe the conservation sentiment in Utah during that decade when it stated "'Conservation' means, if it means anything American, 'wise use.' And 'conservation' means 'water' in this country's semi-arid region."[22]

The vast majority of residents in the vicinity of Dinosaur National Monument viewed it not as a scenic treasure, but as an "area of low economic value with a few remote ranches and a small amount of grazing land which does not exceed $5 an acre in value."[23] It was not very long since western Colorado and eastern Utah had been the edge of the American frontier, and residents of that region still adhered to traditional frontier values. The area was as isolated as any in the continental United States. In 1940, seven of Utah's twenty-nine counties had no bank; eight counties got along without any railroad; and a few counties even lacked a telegraph line.[24] The cost of shipping goods to the area by rail was so high that when residents of Vernal decided to build a bank, they'd actually saved money by having the bricks delivered *through the mail*. Utahns living in such isolated conditions had to be as resourceful as the relatives who had pioneered the land only a generation or two earlier. "The tradition of the pioneer that is strong all through the West," noted one astute observer in 1942, "is a cult in Utah."[25] To Utahns, wilderness was not to be gawked at; it was to be subdued and made productive.

The state's impressive tradition of irrigation only increased the appeal of a water development project at Echo Park. The Mormons who first settled Utah in 1847 took quite literally God's directive to go forth into the desert and make it blossom as the rose. On the day they decided to make a home in Great Basin, the Mormons diverted water from a stream, flooded the land, and put potatoes in the earth. Within three years they had managed to bring water to 16,333 acres of parched land, and that total climbed to an extraordinary 263,473 acres by 1890. Little more than a generation after settling at Salt Lake, the Mormons occupied almost every arable acre for hundreds of miles.[26] By the 1940s, roughly 1.5 million arid acres were being intensively cultivated in Utah. Water storage and diversion had become a fundamental part of life there. "The Utahn who goes to New England or Oregon looks at the broad rivers almost bitterly," acknowledged one Utahn as he attempted to capture the essence of his home. "It is [to the Utahn] unnatural that rivers should waste into the sea, just as it is unnatural that farmers should mature crops by rain alone. Rivers should be damned at canyon mouths and their waters carried in canals to the thirsty land. Water in Utah is precious, savored as champagne might be in another land." Water boosters in Utah considered the state's efforts to capture that resource to be nothing short of "an epic of human ingenuity."[27]

If all this were not enough, religious sentiment in Utah was also predisposed to water development projects; water, after all, would allow the Mormon Saints to prosper. When the Bureau of Reclamation began to promote the Echo Park dam, Utah's population was still overwhelmingly comprised of Mormons who had taken to heart the gospel of desert conquest sanctioned by God. Although there were plenty of Utahns who were not Mormon, the church colored social and cultural norms throughout the state and as far east as Grand Junction, Colorado. While it may seem somewhat cursory to ascribe specific religious values to the population of an entire state, Utahns working on the Writer's Program of the Works Progress Administration admitted as much themselves. "Although the total Church membership numbers perhaps only three-fifths of the population," they asserted in 1940, "the particular quality of Utah life is almost wholly Mormon." And the Mormon social order was built on irrigation. "The Mormon village," one Utahn explained in 1942, "is a green village." To the devout Mormon "it was unthinkable that the gathering place of the Saints should be a barren desert. It should be made to blossom, and it was. . . . Even in the outposts on the edge of the desert . . . there are plantings, there is a quality of order and permanence."[28]

In light of these considerations, it should have been easy to predict the local support for a dam at Dinosaur. One resident of Vernal later confirmed what seems obvious in hindsight. "[Opinion] in Vernal was always to build the dam, right from the beginning," he recalled. "They never did see anything wrong with it." That attitude stemmed at least in part from both the frontier values and religious fervor which permeated Utah. "The mood in Vernal," he said, "was 'here comes a boom—make the desert bloom—we'll all be rich.'"[29]

When World War II ended, the Bureau of Reclamation began to look forward to the enormous appropriations it would need to fully develop the Upper Basin. Residents of the Upper Basin, in turn, began to look forward to receiving their share of the benefits.

Enthusiasm for a federal reclamation project started to grow in communities around Dinosaur as early as March 1946, when residents of Vernal learned that Bureau of Reclamation technicians were surveying a road to the Echo Park site.[30] Before the month was out, the *Vernal Express* was promising residents, "Echo Park Project to Bring Big Industrial Growth." The same paper would later wax poetic about the benefits a dam would bring. "The Uinta basin of Utah and Colorado, adjacent to the Echo Park and Split Mountain dams, has a great power demand of its own. Awaiting a cheap and an abundant power supply are the fabulous natural resources of this area. On the Utah side alone, centering about Vernal are vast deposits of phosphate, gilsonite, asphalt, oil shale, and coal. The Ashley National Forest is humming with sawmills, while oil exploration and drilling are moving on apace. On the Colorado side, oil field, coal, and oil shale make ever greater power demands."[31] The Vernalite who sat on his porch reading the paper could almost hear the wheels of an empire beginning to turn.

Officials from the Bureau of Reclamation were careful to maintain close contact with state officials and local business leaders in Utah and Colorado. Collaboration between the Bureau and the beneficiaries of a major reclamation project was of course necessary to ensure that the project was tailored to the needs of the people it would serve. But this sort of collaboration was not the Bureau's only motive for keeping state officials well informed. All proposed reclamation projects normally had to bear the scrutiny of the Bureau of the Budget before

the commissioner of Reclamation could request authorizing legislation. State officials, however, could petition their congressional delegation for such legislation at any time. By letting state officials take the
lead on the Echo Park project, the Bureau of Reclamation might avoid
the unwelcome attention of tight-fisted economists.

To ensure local support for Echo Park dam, the Bureau tied it to
the Central Utah Project (CUP), one of a dozen "participating projects"
throughout the Upper Basin that would distribute the water made
available by several main-stem storage dams.[32] The reservoirs, canals,
and diversion tunnels included in the CUP would bring water to
150,000 acres of arable but unirrigated Utah farmland, as well as
300,000 acres that was insufficiently watered. The Echo Park dam
would not directly provide any water for irrigation, but it would
replenish Uinta streams that had been diverted to the Bonneville
Basin. The replacement water could alternatively come from a dam at
Flaming Gorge, but Utahns preferred to use Echo Park because it
would store water from the Yampa River. Utah had been promised
water from the Yampa under article 13(a) of the Colorado River Compact, and the Echo Park dam was the only means by which the state
could collect on that promise. Proponents of Echo Park dam would
on that basis begin to tout it as "Utah's last waterhole"—a persistent,
if not wholly accurate, rallying cry.[33] Locals who had little understanding of the technical aspects of the CUP or of the legal tangles
of the Colorado River Compact simply assumed that water for irrigation and municipal use was to come from the Echo Park reservoir. The
Bureau of Reclamation did little to dispel the myth. Instead, it began
to refer to the Echo Park dam as the "wheelhorse" of the entire Upper
Colorado Storage Project because revenues generated by the sale of
the dam's hydroelectricity would finance the construction of projects
elsewhere in the basin. Without the Echo Park unit, Utahns soon
believed, the entire Upper Colorado Storage Project would be in
jeopardy.

Although new Reclamation Commissioner Michael Straus warned
Vernalites that his agency's initial investigations of the dam sites at
Dinosaur were only a preliminary step which in no way ensured the
feasibility of the Echo Park and Split Mountain dams, the disclaimer
was drowned in the enthusiasm his Bureau was generating for the
Central Utah Project.

Until a specific development plan was approved by the secretary
of Interior, however, local boosters for the dam had no real reason
to think a project was imminent. In April 1946, a delegation of ten
Utahns traveled to Washington to meet with Secretary of Interior Julius
Krug and lay the groundwork for the Central Utah Project, including
the Echo Park dam. Krug had been something of a surprise choice

UPPER GREEN RIVER AND TRIBUTARIES

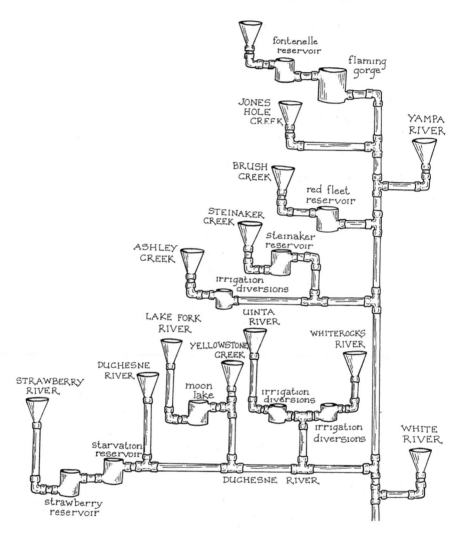

For thousands of years the rivers of the Colorado River System swelled and raced with muddy spring snow melt. As summer progressed, tributaries slacked their flow. The life-giving wild rivers watered the ribbon of vegetation hugging the bank. The rivers were home for water fowl and native fish. Native Americans camped in the shade by the cool water. Today the wild and natural elements are gone, and we have engineered the rivers into a plumbing system for agricultural irrigation, industry, and watering lawns. *Dinosaur National Monument.*

to replace Harold Ickes as secretary of Interior a few months earlier. He was not a westerner, but because he had served as a staff member of the Tennessee Valley Authority and more recently as a member of the War Production Board, most westerners believed he would place emphasis on resource development rather than preservation.[34]

Although no reclamation projects had yet been authorized for the Upper Basin, local officials had good reason to believe that federal money would soon be poured into the area and all were eager to get their fair share. The chairman of the Utah delegation assured the people waiting back home that "we are going to lay our cards on the table and get in our bids for consideration with the rest of the states demanding reclamation funds from the Department of Interior." The implication was that the states which were slow to act would be over-looked when the federal money started showering down on the Upper Basin. While Krug remained noncommittal on the Central Utah Project, Commissioner Straus privately offered the possibility that the Echo Park unit could get advance approval.[35]

Bureau surveys in Dinosaur proceeded slowly, but enthusiasm for the project hardly wavered. In the summer of 1947, the Upper Colorado River Commission finally agreed on a percentage division of the Upper Basin's apportioned water. Under the terms of this agreement, Colorado would receive 51.75 percent of the Upper Colorado River Basin's flow, Utah would receive 23 percent, Wyoming 14 percent, and New Mexico 11.25 percent. In the course of the negotiations, however, the Bureau of Reclamation discovered that it had grossly overestimated the Colorado's average annual flow. The revelation created an even stronger incentive for projects that could regulate flows in the Upper Basin.

The Bureau responded promptly. That same July, it published its two-stage plan for developing water resources of the Upper Basin. At some point during their eight years of investigative studies, Bureau engineers had decided that the Echo Park site was better than the Split Mountain site. Rather than choose between them, however, the Bureau simply decided to build a dam at both sites. The Echo Park dam, as the Bureau now conceived it, would be a 529-foot-high con-crete gravity dam with an installed power capacity of 200,000 kilo-watts. Its reservoir, storing some 6.4-million-acre-feet of water, would extend sixty-three miles up the Green River and forty-four miles up the Yampa. The Split Mountain dam would be smaller, a re-regulating dam to catch peak releases from the Echo Park dam. Although it would also generate some electric power, the Split Mountain dam was not nearly as important as the Echo Park unit, and it could be post-poned until the second phase of the development plan.[36] The prior-ity now was clearly on the Echo Park unit.

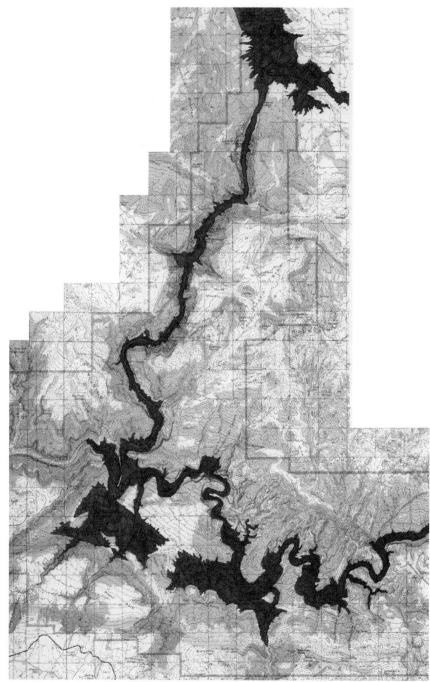

Artist's rendering of flooded areas that would have resulted from construction of the Echo Park dam.

Near the end of 1947, Bureau engineers finally announced to local groups that the topographical studies for Echo Park dam were complete and that studies for the Split Mountain unit were underway. Parley Neeley, the Bureau's regional engineer, also told the Vernal Lion's Club that he predicted the Echo Park dam would be built whether or not the rest of the Central Utah Project won approval. Given these developments, there was every reason for residents around Dinosaur to be optimistic. Within another year, the Upper Colorado River Commission and the Colorado River Basin States Committee jointly resolved that the Echo Park dam begin the initial phase of construction. Prospects for construction of the Echo Park dam looked so good that Vernal dubbed itself "the dam city" and its mayor confidently assured residents that the project "is nearer than we think."[37]

Politicians running for national offices in 1948 knew that water development was one of the most important—perhaps *the* most important—campaign issue in the Upper Basin. Democratic and Republican candidates both gave voters ample reason to be optimistic about the prospects for some comprehensive river development in that region. On one hand, Democrats wanted to build on the New Deal legacy of public power and comprehensive planning in resource development. Harry Truman warned that a fiscally conservative Republican administration would turn power production over to private utilities and turn off the flow of federal reclamation money. On the specific issue of Dinosaur, Truman made his pro-development position clear by stating, "It has always been my opinion that food for coming generations is much more important than bones of the Mesozoic period."[38] If Republicans were not as quick to push the Upper Colorado Storage Project, they at least promised that western states would have greater control of their own local resources. This was an especially appealing message to western businessmen who were trying to open federally reserved lands to mineral exploitation and lumbering and to cattlemen who wanted to end federal supervision of public rangelands. For those voters in Utah and Colorado who wanted economic development in their states and who placed priority on the Dinosaur dams, it must have been difficult to determine which party was promising more.

Although the National Park Service had been slow to respond to the Bureau's power withdrawals at Echo Park and Split Mountain, by the beginning of 1949 it was becoming increasingly hostile to the proposal. When the Bureau of Reclamation conducted its first road surveys to Echo Park, the Park Service responded by proposing ways to preserve scenery along the route, as if the road was a foregone conclusion.[39] Apparently the Park Service strategy was to simply stall for time. "We should protect such authority as we have within the area," Regional Director Lawrence Merriam wrote in a confidential memorandum to

Drury, "until it is necessary for us to relinquish it."[40] The Park Service's "Survey on the Recreational Resources of the Colorado River Basin," completed in 1944, opined that a reclamation project in Dinosaur would constitute "a lamentable intrusion" and would "deplorably alter" the canyons. But the report made no strong objections to reclamation surveys in the monument, stating only, "If and when it is shown that it would be in the greater national interest to develop the water resources of the canyon . . . the status of the unit should be changed to that of a multiple purpose area in which . . . the generation of power would be the principle use, and recreation a secondary but also important use."[41]

When the war ended, Park Service officials began more aggressively to protest water development schemes in the park system. In September 1945, Newton Drury told Ickes that he was "gravely concerned" about the expanding dam construction programs of the Bureau of Reclamation and complained that advanced notification of these projects to state and local officials produced undue pressure on the Park Service. He warned Ickes that "if our park boundaries are adjusted to include these projects, we are only opening the door to other adjustments and further encroachments."[42] Judging from his reaction to the Big Thompson project a few years earlier, it seems likely that Ickes appreciated Drury's appeal to nip development projects in the bud. However, the secretary may have thought Drury's appeal insincere, since Ickes was already angry with him for giving in to pressure to open Olympic National Park to timber interests. Had Ickes known at the time about the secret memorandum of understanding that Drury had signed with Commissioner Page a few years earlier, he almost certainly would have fired the director on the spot.[43]

Drury's concerns notwithstanding, the Park Service had avoided every opportunity to go nose to nose with the Bureau of Reclamation. The Park Service had made some private objections within the Interior Department, but publicly it had not whispered the slightest complaint. Residents of Vernal actually thought that the Park Service *supported* the project since it seemed to be "aware of [the dam's] compensating factors" and to hold "a realistic view of coming events."[44] Secretary Krug may have listened to Drury's appeals, however, for in the spring of 1948 the Park Service received some unexpected help from high officials at the Interior Department. In June, Assistant Secretary William Warne wrote to Straus in regard to investigations underway in the Grand Canyon. "I do not consider it appropriate to investigate any water project that involves a National Park without advance clearance of the Secretary," Warne told the commissioner.[45] Because Warne was himself a strong advocate of water development, it seems likely that the mandate for the memo came directly from Krug. During the summer, Acting Director Arthur Demaray decided Warne's memo could be applied just as validly to Dinosaur and informed the Bureau that the Park Service would no longer permit reclamation surveys within the monument unless it was shown that no other sites were feasible. In August, Demaray personally contacted Commissioner Straus to re-emphasize that in no instance were Bureau surveyors to deface any natural features of the monument and to inform him that when the survey crew left Dinosaur at the end of the season, it would not be permitted to re-enter the area.[46]

The Bureau, however, was not about to give up studies that were so near to completion. Its operations had expanded to the point where they required a specially designed, sixty-foot barge to carry supplies, crews, and drilling rigs up and down the river. Park Service officials at Dinosaur, in contrast, had no money to purchase any boats of their own and were forced to hire Vernal riverman Bus Hatch if they wanted to float the river in the monument.[47] When the summer came to a close, the Bureau's regional engineer informed the Park Service that he expected the work "season" to last through the winter, until the following May or June. In the meantime, local enthusiasm for the project continued to snowball. By the end of the summer, even the most optimistic regional Park Service officials were ready to acknowledge that they "might well be giving some thought to the possibility" of the dam's proponents "being successful in abolishing that part of the monument affected by the reclamation activities," unaware that their own director had secretly agreed to this course of action more than seven years earlier.[48]

By the end of the summer of 1949, most residents of the Uinta Basin would have recognized the Bureau of Reclamation's insignia. The Bureau had been highly visible in the area since its technicians had arrived at Split Mountain in 1939, and it had gone to great lengths to drum up support for a local reclamation project. Accordingly, expectations were running high when Secretary of Interior Julius Krug arrived to make an address in Glenwood Springs, Colorado, in September of 1949. Although Krug had up to this point refused to commit himself to the dam at Echo Park, conventional wisdom in Utah suggested that the long wait was almost over. Many of the dam's proponents no doubt believed that Krug was arriving in Glenwood Springs to announce in dramatic fashion his support of the Echo Park project and to say some inspiring words about the economic boom that would follow its construction.

The next day, however, the *Vernal Express* displayed a shocking headline: "Krug Turns Thumbs Down on Echo Park Project." In reference to the headline the paper later noted, "Only a black border on the paper could have made it a more doleful edition." It is not clear why Krug refused to support the project, or why he chose to reveal his decision while in Glenwood Springs. But the secretary's rationale as he officially stated it was exceedingly clear. "Large power and flood control projects should not be recommended for construction in national parks," Krug said, "unless the need for such projects is so pressing that the economic stability of our country, or its existence, would be endangered without them."[49]

Dumbfounded by Krug's decision, the dam's boosters turned to the Bureau of Reclamation and asked for an explanation. Red-faced Bureau officials could only answer by pointing at the Park Service. Up until that point, local residents had no reason to suspect that the Park Service was opposed to the dam. In light of the assurances that had been made when the monument was enlarged in 1938, any opposition by Park Service seemed to constitute an unforgivable breach of faith. In Vernal, hostility against the Park Service flared. By the end of September, the superintendent of Dinosaur concluded that Krug "might have kicked the pot over" in making his announcement and could easily surmise that "the National Park Service is not the most popular bureau of the federal government in and around Vernal at this time."[50] The trouble further fueled the animosity between the Park Service and the Bureau of Reclamation because the Park Service had until this point managed to keep what it knew to be an unpopular position wholly within the Interior Department. Park Service officials in the regional office suspected the Bureau was blaming the Park Service for Krug's decision and was going out of its way to arouse public enmity. "We are at a loss to know just how this information on the Service's position has become known outside the Department, since [it] was not to be released outside the Department at this time," the regional director fumed in a memo. He added hotly, "It has been a little embarrassing to find that what we thought was confidential within the Department was not so in fact."[51]

Despite the growing animosity in the Department of Interior, the Bureau apparently felt no obligation to refrain from promoting the Echo Park project. The Upper Colorado River Commission and the Colorado River Basin States Committee had already reviewed the Bureau's data and recommended that the Echo Park dam be placed in high priority. In early October, at a meeting held by the Colorado Water Conservation Board, the Bureau of Reclamation, and other interested water users, a resolution passed favoring the Echo Park and Split Mountain dams and an appropriate adjustment of Dinosaur's land status. A month later, a commission of five Upper Basin states approved a development plan that included both the Echo Park and Split Mountain units.[52]

As the Bureau of Reclamation marshalled increasing local support for the dams, the Park Service wavered. In October, Drury privately noted that his agency's approach to the Bureau of Reclamation had to be "gauged by the realization that it is now the eleventh hour in the planning phase of these projects," and he conceded that "the Echo Park project is of the highest priority in the Upper Basin projects, with strong public support, whereas Dinosaur National Monument has not been made accessible to the public and the canyon portions of it are virtually unknown." Drury clearly felt he was losing the battle and thought it "advisable to indicate the direction in which a possible solution might be reached that would minimize the damage to the monument." Drury indicated that the situation was so desperate that he would be willing to allow a dam at an alternate site in the monument so long as Echo Park was spared, and in November, he officially made this proposal to the secretary of Interior. Soon after, the National Park Service granted permission for the Bureau to conduct triangulation, topography, and road surveys, effectively retracting its four-month-old policy of refusing permission for further studies in Dinosaur.[53]

By the time Oscar Chapman replaced Julius Krug as secretary of Interior at the end of 1949, it was clear that, barring an extraordinary turn of events, the Bureau of Reclamation would build its dams in Dinosaur National Monument. The Park Service's position on the dams had been reactionary from the very beginning. To be sure, the Bureau had a commanding political advantage over the Park Service, both in terms of appropriations and highly vocal constituent support. The Bureau also controlled the generation and dissemination of the technical information which ultimately would be used to judge the suitability of the Echo Park and Split Mountain dam sites. But the Park Service made several miscalculations that increased the Bureau's already considerable advantage. Assurances that the Park Service would not interfere with future power development when the monument was enlarged were probably based on the false belief that no such development would be needed. Confidence that public opinion would protect the monument failed to take into account the economic and religious values that shaped local reaction to the Dinosaur dams. Moreover, the Park Service made few attempts to oppose the Bureau's preliminary surveys in the monument until 1949, when they were nearly complete, even though Secretary of Interior Harold Ickes unabashedly favored the Park Service over the Bureau of Reclamation and told Newton Drury as early as 1945, "I think we ought to go very slowly about giving the . . . right to bore in any of our parks."[54] The Park Service director would later reflect on his agency's reaction to the dam proposal and say, "I didn't think we should compromise at all." But Drury himself seems to have been the first to give up the fight. "Dinosaur," he concluded, "is a dead duck."[55]

CHAPTER THREE

A House Divided

O n the morning of April 3, 1950, Dr. Ira N. Gabrielson walked past
the cherry blossoms in Rawlins Park and up to the doors of the
Department of Interior building in Washington, D.C. As he proceeded
to the auditorium where Secretary of Interior Oscar Chapman would
soon open a hearing on the proposed Echo Park dam, he passed several murals commissioned by former Secretary of Interior Harold Ickes.
In the auditorium itself—an enormous, two-story room—hung perhaps
the most impressive piece of artwork, a bas-relief illustrating John Wesley Powell's exploration of the Colorado River. Gabrielson must have
thought it appropriate that Powell would be looking down on the hearing, for the one-armed explorer embodied much of the ambiguity and
ambivalence of water management in the Colorado River Basin. As one
of the first people to recognize the importance of water in the arid
West, Powell was an unabashed advocate of reclamation and irrigation.
But he also warned his contemporaries that the agricultural opportunities in the arid regions of the West were limited, and he suggested
that the government take an active role in western water development
to prevent the misuse of that vital resource. Proponents of the Echo
Park dam no doubt admired Powell's enthusiasm for water development. Conservationists hailed him as a visionary who recognized that
development in the West should be limited by ecological realities. In
a sense, the hearing was intended to help resolve that ambivalence.[1]

Only a few weeks earlier, National Park Service Director Newton
Drury had asked Gabrielson to appear at the hearing, to prepare a
statement in favor of preserving Dinosaur National Monument, and

Three secretaries of Interior:
Oscar L. Chapman (top left),
Julius Krug (top right), and
Harold Ickes (bottom).
*National Park Service,
Historic Photograph Collection.*

to provide whatever leadership he could for the assemblage of witnesses who would testify against the Echo Park dam. As president of
the Wildlife Management Institute, and former director of the U.S. Fish
and Wildlife Service, Gabrielson enjoyed considerable influence with
conservationists both in the government and in private organizations.
Representatives from many of these conservation organizations had
come to the hearing to join Gabrielson in testifying against the Echo
Park dam, including Waldo Leland of the National Parks Advisory

National Park Service Directors Newton Drury (left) and Stephen T. Mather (right). *National Park Service, Historic Photograph Collection.*

Board, William Voigt of the Izaak Walton League, Kenneth Morrison of the National Audubon Society, Fred Packard of the National Parks Association, Bestor Robinson of the Sierra Club, and Howard Zahniser of the Wilderness Society. If Gabrielson was impressed by the collection of conservation leaders that had been assembled for this hearing, however, he must have been equally struck by the luminaries who had turned out to testify in favor of the Echo Park dam. Four western senators and five congressmen were in the room, in addition to a cadre of delegates representing the business and municipal interests of the Upper Basin states and a whole support team of engineers from the Bureau of Reclamation.

Newton Drury, of course, was also present, and he must have been pleased with the number of conservationists who had shown up on short notice. Drury knew his agency was behind on points as it entered the final rounds of its fight with the Bureau of Reclamation, but he was hopeful that the testimony of these conservationists would allow him to score a knockout punch. The Bureau had benefited greatly from the regional support for the Echo Park dam in the Upper Basin states, and Drury was hoping that these private conservation organizations would at long last give his agency its own constituent voice.

This reliance on private organizations to defend the intangible, aesthetic values of the park system was nothing new. Only three years after becoming the Park Service's first director, Stephen Mather spent much of his own money to establish the National Parks Association as a private watchdog organization, because he felt that only a private organization could "defend the National Parks and the National Monuments fearlessly against the assaults of private interests and aggressive commercialism."[2] A number of private organizations, including the Izaak Walton League and the Sierra Club, had subsequently helped Mather keep irrigationists out of Yellowstone National Park in the 1920s. Unlike the various federal agencies which had been established to manage resources, private conservation organizations had no financial or political interests to protect. They therefore tended to place more emphasis on the intangible values of nature, including its value as wilderness. Drury knew the national sentiment for wilderness preservation was strong and he could take comfort in the fact that the conservationists assembled in the auditorium were among the most qualified people in the country to give voice to that sentiment.

Shortly after 10:00 AM, the auditorium fell silent as Secretary of Interior Oscar Chapman stood to make some opening remarks. Everyone present must have been aware of the hot seat in which Chapman was sitting. While serving as assistant secretary under Harold Ickes and Julius Krug, Chapman had witnessed the growing conflict between the Bureau of Reclamation and National Park Service from a position of relative neutrality. When President Truman nominated Chapman to replace Krug, advocates of wilderness preservation and water development were both cautiously optimistic. Ickes spoke for many conservationists when he proclaimed that "President Truman could not have done better" in choosing Chapman.[3] But most proponents of an Upper Colorado Storage Project believed that Chapman, a Coloradan, would be sympathetic to the cause of water development. The advocates of the project certainly must have been pleased with the definition of conservation Chapman gave at his confirmation hearing. "[It] does not mean, as many of our people are prone to think, the locking up of some resources in order to keep people from touching or using it," Chapman stated. "It means to develop the resources in a wise way." Yet when Utah Senator Arthur Watkins questioned him about the Upper Colorado Basin project, Chapman tried to maintain his neutrality by stating that each reclamation project had to be evaluated on its own merits. When Watkins pressed him on the Echo Park dam, Chapman remained noncommittal. "I could not give you a decision on that this morning," Chapman told the senator with a smile, ". . . if my nomination depended on it."[4]

Upon taking the office of secretary, however, Chapman found the Dinosaur conflict sitting squarely in his lap. In his earliest statements on the issue, Chapman tried to express his ambivalent feelings about the Dinosaur dams. "I accept fully and unreservedly the idea, now long established, that there rests on us the obligation to set aside and protect areas of superlative natural scenery," Chapman told a meeting of the American Planning and Civic Association. "The long term national interest must govern the decision on any proposal that would destroy, impair, or even modify any part of the . . . national park system." Dinosaur, certainly, was the most dramatic case in point, but Chapman was careful to add that "as enthusiasts for sound conservation measures we cannot fail to recognize that the needs of a rising population and an expanding economy are giving increased importance to our programs for development and utilization of the nation's limited water and other natural resources."[5]

More than at any other time in the past, the Interior Department was a house divided. Because it was responsible for the administration of the nation's vast public domain, the Interior Department had always attracted people who claimed to be "conservationists" in some sense of the word. Yet among the Interior's various agencies, there was widespread disagreement about the meaning of conservation and how conservation principles should be applied in specific situations. Some, especially in the Park Service, felt that preservation of wilderness was at the heart of conservation. Others, especially in the Bureau of Reclamation, saw conservation as the wise use of resources. That two agencies in his department, both full of so-called "conservationists," could pursue such sharply conflicting policies made the secretary of Interior's position an uncomfortable one. Harold Ickes had been well aware of this ideological division in his department and had tried to mitigate the conflict by swapping the Bureau of Reclamation for the Department of Agriculture's Forest Service, thereby creating a more consolidated "Department of Conservation." Chapman likewise recognized the Interior's "dual responsibility—to protect and to develop our nation's resources." By the time Chapman took office in December 1949, Interior encompassed such a diverse spectrum of conservation philosophies that it may have seemed as if John Calhoun's cynical, century-old prophesy that "everything on the face of God's earth will go into the Home Department" had indeed come true.[6]

Within weeks of Chapman's confirmation, National Park Service Director Newton Drury and Reclamation Commissioner Michael Straus each began clamoring to win the ear of their new boss. After receiving conflicting recommendations on the Echo Park dam from the two agencies, Chapman decided in February to postpone any decision until he had time to study the proposal in greater depth. Knowing that Chapman was genuinely undecided on the issue, both Straus and Drury decided to present their case in more detail. In a confidential memorandum to the secretary, Straus wrote, "In my opinion, the record is clear that despite any legal interpretations of the proclamation, it was the intent of the department and of the president to reserve the water power potentialities of the Green and Yampa rivers within Dinosaur National Monument. It is the intent to which we must look and not the exact language." To Straus, the government's intention was clear, and in his attempt to persuade Chapman, the Commissioner Straus cited the Federal Power Commission's decision to explicitly reserve the right to develop a power project at Dinosaur. The commission's desire to reserve "two units in the center of the scheme," Straus noted, could refer only to the Echo Park and Split Mountain sites.[7]

Drury similarly implored Chapman to look at the intention of the law. "The review and clearance procedures which Reclamation now wants rushed through will comply only with the letter of the law," Drury claimed. "The spirit has already been violated through the irregular, advanced submission by the Bureau, to the upper Colorado River basin states concerned, of its preliminary draft report." The Bureau's disregard for proper departmental procedure had created a disparity of information, and Drury found it "regrettable that the Department should now be embarrassed by the resulting local 'demand' for the Echo Park and Split Mountain proposals in advance of secretarial determination of the Department's position." Drury was also quick to point out the restraint of his own agency, noting that the Park Service had "refrained from informing the conservation organizations and others having an interest in this specific problem." The disparity in information put the Park Service at a severe disadvantage, and Drury recommended that Chapman "have their views before reaching a conclusion on this controversial issue." In March, Drury again expressed his "hope that some arrangement can be made for future handling of projects such as this one . . . whereby all interested parties—federal, state, and individual, including the many conservation societies—can be simultaneously advised and brought in for general discussion at one time, especially before approval of any one particular interested group is obtained."[8]

The secretary apparently was swayed by Drury's appeal. Soon after receiving Drury's memo, Chapman arranged for a special hearing on the Echo Park project to take place early in April. Although he knew that conservation organizations would be hard pressed to prepare for the hearing on such short notice, Chapman was under considerable pressure to settle the matter as soon as possible. His three-month silence on the issue had already begun to annoy the dam's proponents, among them powerful senators like Arthur Watkins of Utah, who wanted immediate authorization of the Upper Colorado Storage Project. California was already using all 4.4-million-acre-feet of water allotted to it by the Colorado River Compact and planning a second aqueduct to Los Angeles that would divert an additional 700,000-acre-feet of "surplus" flows. Although the Upper Basin states couldn't use these flows, at the time they worried that if California began to borrow this water, the Upper Basin would have a hard time getting it back.[9] B. H. Stringham, speaking for the Vernal Chamber of Commerce, captured the mood of many in the Upper Basin when he stated, "We are fighting for our last waterhole, for without Echo Park dam there will be no Central Utah Project, and unless we are able to build the Central Utah there is no other way in which we of Utah can ever use our rightful share of the Colorado River."[10] The Bureau of Reclamation also had reason to push for a quick authorization, as it feared that private power companies might come into the area and absorb most of the market for electric power. By the end of January, the *Vernal Express* warned residents that the Echo Park project was "imperiled by bureaucratic delay" and urged them to flood Chapman's office with letters in support of the dam. Fifty-six local chambers of commerce in Colorado and Utah submitted resolutions to let the secretary know the dam was needed, and now.[11]

Chapman, however, was determined to weigh national interests against the regional demand for the project so that his decision would serve "the greatest good for the greatest number of people." The public hearing would be the means of weighing those interests. As Chapman stood to open the hearing, he assured all present that he would "sincerely and honestly" consider all testimony before making a final decision.[12]

After brief opening statements by both sides, the dam's proponents were given the floor to make their case. Officials from the Bureau of Reclamation attempted to illustrate the importance of the Echo Park and Split Mountain dams to the Upper Colorado Storage Project. The Upper Basin, they noted, had a much more erratic supply of water than the Lower Basin. The storage project, which would include anywhere between four to eight additional main-stem storage dams, was intended primarily "to regulate the 'peaks' and 'valleys' in the river flow" so that the Upper Basin would have enough water to fulfill the terms of the Colorado River Compact. To Bureau engineers, it was "obvious that considerable storage capacity is necessary to permit carrying water over from the wet years to the dry years"—some 48,000,000-acre-feet of storage, according to Bureau estimates. The Echo Park dam was "one of the very few sites which can accomplish such control in the upper reaches of the basin to any large degree," and therefore it was indispensable to the project. Finding a substitute site might be technologically possible, Bureau engineers admitted, but none would be as good as the Echo Park site. "Any group of reservoirs which does not include Echo Park and Split Mountain can meet the objectives heretofore outlined only at the cost of increased evaporation loss, less annual revenues, and higher unit power costs," the Bureau emphasized. Any alternative site would increase evaporation rates by some 200,000- to 350,000-acre-feet per year—enough water to serve a city the size of Denver. In addition, Echo Park and Split Mountain had "very attractive power drops . . . from which the deficient market of major load centers in Utah, Colorado, and Wyoming can conveniently be served with hydroelectric energy." The Bureau concluded that "any substitution would entail a serious waste of the water and power resources to the detriment of the Nation and particularly to the people of the Upper Colorado River Basin."[13]

Anticipating the arguments of the conservationists at the hearing, the dam proponents also claimed that the reservoir would "greatly enhance" Dinosaur National Monument. "Present visits to the monument . . . are limited essentially to the museum and quarries located at the original 80-acre site," the Bureau of Reclamation noted. "Views of the canyon sections of the monument have been appreciated by a mere handful of hardy river runners who on infrequent occasions have made the hazardous boat trip through the rapids." The dam at Echo Park would change all that. By filling the canyons with still water, the dam would make the monument much more accessible to the public. "These scenic gorges and primitive areas were withdrawn for the recreation of the people," the Bureau pointed out, and no true conservationist would oppose a project that would make them easier and safer to reach.[14]

Finally, delegates from the Upper Basin states appealed to Secretary Chapman's sense of justice. Several witnesses testified that the Park Service had promised not to interfere with power development when Dinosaur was enlarged, and one Park Service official signed an affidavit in which he admitted that such assurances had been given. Chapman probably allowed such arguments to carry considerable weight because shortly before the hearing Straus had presented him with the "memorandum of understanding" signed by Drury and Reclamation Commissioner Page back in November 1941. Chapman chose not to reveal the document at the hearing for fear of embarrassing Drury and the department, but he would later consult with former Secretary Harold Ickes, who reportedly was furious that such an agreement was made without his approval.[15]

The testimony in favor of Echo Park dam lasted the entire morning and well into the afternoon. Ira Gabrielson, the first person to testify against the dam, must have been discouraged that the Bureau engineers had so much technical information to support their position and that they had been given so much time to make their argument. Gabrielson tried to rally his colleagues. "It was our understanding that [conservationists] would be given equal time with the proponents of this proposition," Gabrielson stated, "but it is already plainly evident that we will not. . . . I think it is only fair to say that we are disappointed at the present time in the way these hearings have been handled." Having boldly taken the secretary to task, Gabrielson proceeded to lay out the argument against the dam. "Conservationists are weighing, and are asking you to weigh . . . other values which cannot be appraised in dollars and cents, even though in the long run they are of tremendous importance to the economy of the surrounding region and to the people of the nation."[16]

A number of witnesses attempted to describe these intangible values. "The wilderness value," stated George Kelley of Colorado, "is something hard to define and measure . . . but it is, nevertheless, a real value and a value that is rapidly being lost in our country." The president of the American Planning and Civic Association declared, "Our industrial civilization is creating an ever greater need for the average man, woman, and child to re-establish contact with nature, to be inspired by and appreciate the wonders of nature, and to be diverted from the whirling wheels of machinery and chance." Kenneth Morrison of the National Audubon Society backed him up. "Even a few days in places where canyons rise sharply overhead, swift waters tumble at their feet, and bird songs blend with wind through trees, sends wilderness travelers back home with renewed mental and physical vigor." Morrison added almost as an afterthought, "No one has ever been able to place a dollar sign on wilderness values. Who can say what the value of an unspoiled Dinosaur National Monument will be in an era when the face of the nation has been almost completely irrigated, drained, and dammed?" Another conservationist pointed out that even "from a chamber of commerce viewpoint" the dams did not make sense. "The greatest profit to the surrounding communities will come, in the long run, from leaving this area to be developed as a scenic attraction . . . A good deal of our economic support in the future is going to be from supplying recreation to the rest of the United States. It would be short-sighted not to consider that angle." Howard Zahniser rested the conservationists' case on the well-known words of Aldo Leopold. "Conservation," he stated, "must spring from a conviction of what is esthetically right as well as what is economically expedient."[17]

Zahniser and his colleagues made it abundantly clear that they did not consider dams in Dinosaur National Monument to be "esthetically right." Landscape architect Frederick Law Olmstead, Jr., testified that the damage to scenic values caused by the Echo Park project would be "catastrophically great." Conservationists were particularly concerned about the claim that Echo Park and Split Mountain dams would enhance the monument. "Flowing water is an essential part of an exhibit of the Colorado River. . . . Conservationists do not subscribe, and probably never will subscribe, to this specious argument that the stinking debris left by a fluctuating reservoir adds to the beauty of

An aerial view of the Yampa Canyon country. *Photo by Jack Boucher, Dinosaur National Monument.*

any natural scenery," Gabrielson snorted. Harold Ickes could not attend the hearing, but in a written statement he reverently noted that "the occasions have been rare indeed when the hand of man . . . has been able to improve upon God's handiwork. . . . The greatest sculptor that ever lived cannot improve a mountain peak by drill and chisel."[18]

Although they spent considerable time describing the beauty and scenic qualities of the Green and Yampa rivers, conservationists were not concerned solely with the fate of those two rivers. They had come to fight for a *principle.* "In making this plea for the preservation of the Dinosaur National Monument in its natural state, we are also pleading against setting a precedent," stated General U.S. Grant III. "Such a precedent, if allowed, will naturally and necessarily be emulated over and over again, whenever anyone finds some possible profit in the similar desecration and injury to other National Parks and Monuments." Harold Ickes agreed. "Once a national park has been created, they [*sic*] should be held inviolate," Ickes declared in his prepared statement. "To violate one monument is to invite violation of others. We must mean what we say when we declare that any area is to be dedicated to future generations."

William Voigt expressed much the same opinion for the Izaak Walton League. "Perhaps we of the League would not be so disturbed if this were an isolated case, without parallels or extensions. But that is not so. This is but one of many proposals, and the sum total of them, and the successors they most assuredly would have, can only mean the destruction of our national system of parks, monuments, established wilderness areas, and wildlife refuges."

Indeed, there were a number of other dam proposals pending at that moment, including projects that would flood Glacier, Mammoth Cave, Kings Canyon national parks, and Grand Canyon National Monument. More than a dozen units of the park system were seriously threatened by development proposals, and conservationists believed that the very concept of the national park system was in jeopardy. "The matter of precedent is all-important," declared one conservationist. "There is so much pressure to invade the National Parks and Monuments that any concession . . . will be a deplorable retreat in the national policy of conservation."[19]

Because the principle of park inviolability was their primary concern, conservationists took great pains to point out that they did not oppose reasonable development of water resources. "It would be . . . silly . . . for us to claim that all of the scenic areas in the United States . . . should be preserved without encroachment," stated Horace Albright. "We do not take that stand." Like his colleagues, Albright was willing to support other dams in other areas, no matter how scenic, as long as those areas had not been designated as federal reserves. "It is not a question of denying development to these people," noted Gabrielson. "It is a question of whether sufficient effort has been made to find a real substitute." A representative from the Sierra Club admitted that he owned an irrigated ranch and asserted, "I am in thorough accord with the basic conception that our rivers must be developed for power and water, so far as is economically feasible and not inconsistent with more important national policies." Joe Penfold, western representative of the Izaak Walton League, agreed. "We recognize thoroughly the importance of water," he stated. "No one in his right mind can be opposed to the sound and logical development of that prime resource."[20]

Conservationists expected to be pejoratively branded as "nature lovers" and therefore made every attempt to sound reasonable. They even went so far as to offer an alternative development plan. General U.S. Grant III, an engineer of forty-three years experience with the Army Corps of Engineers, pointed out that water and power development could proceed apace without the Echo Park and Split Mountain dams. The notion that sufficient water or power development would not be possible without these two dams was a "very erroneous impression," according to Grant, who claimed that the whole project could be developed "with a slight change in order and a substitution . . . of other projects for these two projects which will answer the purpose just about as well."

Grant proceeded to lay out an alternative plan that he claimed would result in greater storage capacity and power potential than the Echo Park and Split Mountain dams. In return, conservationists demanded that the secretary be reasonable as well. "This is a very tense decision which must be made," stated Howard Zahniser, "particularly if there is a tendency to decide in favor of the construction of this dam. The canyon which we are preserving we can be sure will endure, but the dam site, when once used, will have gone. If we construct this dam, we can no longer think of saving the monument as it is. Yet if we save the monument now and postpone decision until further study, we preserve our choice."[21]

The hearing dragged on late into the night. Chapman left early, but every witness was allowed to testify for the record. By the time each side had finished its closing statement, it was nearly midnight. Exhausted, the representatives withdrew from the auditorium and shuffled out into the cool night air. Drury, for one, was satisfied that both sides of the controversy had been presented fully. The arguments against the dam, he believed, had been delivered "with extraordinary competence, force, and clarity."[22] The only thing for the Park Service director to do now was await a decision. The lines had been drawn.

CHAPTER FOUR

The Strength of Convictions

With the administrative hearings at the Department of Interior behind them, the proponents of the Echo Park dam waited expectantly for Secretary Chapman to recommend authorization of the Upper Colorado River Storage Project. The *Vernal Express* reported that local delegates returned home from the hearings in an "extremely optimistic frame of mind," their spirits especially bolstered by the witnesses who had testified that Park Service officials had pledged not to interfere with water development in the enlarged monument.[1] Despite their bravado, however, Utahns were clearly anxious about Chapman's decision. After several weeks went by without any word from Interior, the *Vernal Express* printed a bold headline declaring, "Chamber of Commerce Wants Yes or No From Chapman." Meanwhile, B. H. Stringham, chairman of the chamber's Echo Park Dam Committee, fretted that "continued delay affords more and more time for the nature-lover groups to apply . . . pressure on Chapman" and "plays California's game."[2] As far as local boosters were concerned, the project was already way behind schedule.

Conservationists also worked to influence Chapman's decision. Drury wrote on behalf of the Park Service "to express its deep gratitude to all those who spoke or wrote in defense of the integrity of the national park system."[3] The Park Service now had what it lacked in the first ten years of the Echo Park debate—a strong constituent voice. Even so, a couple of weeks after the hearing, the National Parks Advisory Board predicted that Chapman would approve the dam. Bernard DeVoto, a Pulitzer Prize winning historian and columnist for *Harper's* magazine, held an honorary position on that board. Because DeVoto, a native of Utah, enjoyed a national reputation as one of the most well-informed critics of western resource policy, the advisory board suggested that he publish an article on the controversy before Chapman announced his decision.[4] DeVoto tried his best, but could not get the article printed in time.

On June 28, 1950, just a few days after United States troops were committed to Korea, the *Salt Lake Tribune* announced the good news to the state of Utah in a bold headline: "Echo Park Dam Gets Approval."[5] Chapman, hoping that he could now put this decisive issue behind him, wrote a joint memorandum to Drury and Straus to explain his decision. "The issue that confronts me is a difficult one," Chapman stated. "I believe that when the Dinosaur National Monument was enlarged the problem of this development was inevitable, but in view of the demonstrated need for water there is only one action I can take. Without intending to establish a precedent by tampering with the violability of our national parks and monuments, I have approved the plan called upon by the Bureau of Reclamation to draft necessary recommendations to the Congress for the building of Echo Park and Split Mountain dams."[6] Chapman also tried to placate the conservationists who would be bitterly disappointed with his decision. "I'm convinced that the growth and development of the West depends upon the adoption of a sound Upper Colorado River Basin Program, and that is the most important consideration to be faced in this matter," he wrote to one. "The decision to recommend construction of Echo Park and Split Mountain Dams should not be construed in any way as relaxing our standards for the preservation of our heritage of natural wonders."[7]

The dam's proponents must have a taken a collective sigh of relief. With executive approval now out of the way, the project faced much better chances for authorization in Congress. Commissioner Straus, eager to move the project from the administrative to the legislative arena, predicted that enabling legislation for the proposed dams would be introduced early in 1951. Officials in Vernal, barely able to contain their excitement over the project, officially designating the second day of September as "Uinta Basin Dam Day," giving themselves an opportunity to bask in the glow of their recent victory.[8]

To conservationists, however, it must have seemed difficult to reconcile Chapman's emphasis on efficient development with his claim that the Dinosaur dams would not set a bad precedent. Those who were sympathetic to Chapman's dilemma suggested that he must have been "stampeded" into approving the project in order to help gain support for the president's foreign policy actions in Korea. Arthur Carhart, who enjoyed a friendly relationship with Chapman, wrote,

"With the planning of several years in front of you, the assembly of many figures and arguments, with proponents for the building of the Dinosaur Dams well-regimented, I can see how cold-blooded weighing of what you had before you brought your decision."[9] Others were less sympathetic. DeVoto thought Chapman unnecessarily bowed "to the pressure of the Bureau . . . and the western block in Congress." Ickes implied the same and stated that Chapman's decision was "a powerful argument in favor of the theory I have had for many years . . . that, other things being equal, a western man should not be appointed Secretary of the Interior."[10]

In fact, Chapman's decision may have been politically motivated. Democratic Senator Elbert Thomas of Utah was facing a tough bid for re-election when Assistant Secretary of Labor John Gibson informed Chapman, "The people in Utah feel that the approval of the project by your Department, tying in Senator Thomas, would be most helpful politically."[11] Most conservationists felt that such political considerations unduly influenced Chapman's decision on the Dinosaur dams. William Voigt of the Izaak Walton League considered Chapman "a smoothie" and noted that "when politics comes in the door, conservation goes out the window. Political advantage carries tremendous weight with Oscar."[12] Such criticism may have been true, but it probably was an unfair indictment of Chapman's commitment to the national park system. The position of secretary of Interior is, after all, a political appointment and Chapman's decision was certainly consistent with the development emphasis of the Truman administration.

Conservationists still believed they could defeat the Echo Park project before it won congressional authorization if they could bring the controversy before a national audience to offset the localized western pressures that had forced Chapman into submission. Keeping the national park system inviolate seemed to be a dramatically simple issue around which all conservation organizations could rally. The opening volley in their anti-dam campaign came four weeks after Chapman's decision with the publication of Bernard DeVoto's article in the *Saturday Evening Post*. With a circulation of over four million, the *Post* enjoyed considerably more readers than *Harper's*, and DeVoto felt it was imperative to bring the controversy to as large an audience as possible. "Shall We Let Them Ruin Our National Parks?" drew nationwide attention specifically to the Echo Park dam and generally to the threat of water and power development in the park system. DeVoto used stinging sarcasm to simultaneously praise the uniqueness of Dinosaur's river canyons and indict the Bureau of Reclamation for wanting to develop them. "If it is able to force the Echo Park project through, the Bureau of Reclamation will build some fine highways along the reservoirs," DeVoto noted wryly. "Anyone who travels the 2000 miles from New York City—or 1200 from Galveston or 1000 from Seattle—will no doubt enjoy driving along these roads. He can also do still-water fishing where, before the bureau took benevolent thought of him, he could only do white-water fishing, and he can go boating or sailing on the reservoirs that have obliterated the scenery. . . . But the New Yorker can go motoring along the Palisades, boating in Central Park, or sailing at Larchmont and fishing at many places within an hour of George Washington Bridge. . . . The only reason why anyone would ever go to Dinosaur National Monument is to see what the Bureau of Reclamation proposes to destroy."[13]

DeVoto's call to action resounded even louder when *Reader's Digest* carried a condensed version of his article in November. A rash of articles on the Dinosaur dams simultaneously appeared in conservation bulletins such as *National Parks Magazine, Audubon Magazine,* and *The Living Wilderness,* but these articles reached an audience that was for the most part already committed to the idea of wilderness preservation.[14] DeVoto's article was certainly the most widely read, and it brought appreciative comments from conservationists everywhere. Horace Albright praised DeVoto for writing "one of the finest national-park articles that has ever appeared," an article which "opens up the fight in a big way and gives strength and encouragement to conservationists everywhere."[15]

Westerners, on the other hand, roundly denounced the article and its presumptuous author. Utahns must have been especially displeased with DeVoto, a native of that state. The *Deseret News* of Salt Lake City called the article "foully misleading" and dismissed it as "a Park Service 'plant' in its intra-mural fight against the Reclamation Service." The *Denver Post* editorialized that the western states should be allowed to do as they pleased with their scenery. Reclamation Commissioner Michael Straus, with a sarcasm of his own, later denounced DeVoto in front of a western audience as one of "the self-constituted, long-distance protectors of Dinosaur National Monument."

"[From] their air conditioned caves overlooking the undeveloped wilderness areas of Central Park in New York, Lincoln Park in Chicago, and Boston Commons in the adopted city of a transplanted western writer who has a tendency to forget his heritage, these self-appointed guardians . . . contend . . . that the highest use for your area and resources is as a museum and cemetery for dinosaur bones," Straus declared.[16] Although he probably bristled at Straus's intentionally misleading suggestion that the 211,000 acres of wilderness at Dinosaur were merely a "cemetery for dinosaur bones," DeVoto must have been pleased to be lumped with such respected conservationists as John Baker, based in New York with the National Audubon Society, and William Voigt of the Izaak Walton League in Chicago.

DeVoto had long felt that the western states were working against their own interests when they tried to end federal management of western resources.[17] He responded to his western critics in his characteristically ironic manner. "The National parks and monuments happen not to be *your* scenery," he wrote in a letter to the editor of the *Denver Post.* "They are *our* scenery. They do not belong to Colorado or to the West, they belong to the people of the United States, including the miserable unfortunates who have to live east of the Allegheny hillocks." He ended with what might have sounded like a threat: "Podner, as one Westerner to another, let me give you one small piece of advice before you start shooting again. Don't snoot those unfortunates too loudly or obnoxiously. You might make them so mad that they would stop paying for your water development."[18]

This statement perhaps as well as any other underscores the impact that DeVoto had on the Echo Park dam controversy. His article expanded the debate from a regional to a national arena. After his article was circulated, the Dinosaur dams were no longer of concern solely to the Upper Basin farmers who would receive irrigation water and the Upper Basin industries which would receive cheap electricity. They were now a concern to every American citizen who held a share in the national park system and every taxpayer who helped to pay for federal reclamation. Although these groups—for the most part easterners—were more diffuse and more difficult to mobilize than the dam's boosters, they clearly represented a tremendous source of political support for the conservation organizations.

During the last months of 1950, the Echo Park controversy subsided as the Bureau's report on the Colorado River Storage Project and Participating Projects was circulated for review by other federal agencies and by the Upper Basin states. At the beginning of January, Utah Senator Arthur Watkins announced that he would introduce legislation to authorize construction of the Echo Park and Split Mountain dams at the opening of the 82nd Congress. When the first session opened, a number of bills were introduced to authorize the initial phase of construction.[19] However, because the federal review process was not yet complete, these bills drew little attention and no action was taken.

In February 1951, the simmering tension in the Department of Interior erupted onto the public scene, and the Echo Park dam was again a topic for national media attention. Since issuing his decision on the project, Chapman had been trying to mend the intra-agency rift in his department. To that end, he had instructed Newton Drury to restrict his agency's communications with the private conservation organizations which were protesting the decision.[20] Drury did not want to lose the constituent voice that had finally put his agency on equal footing with the Bureau of Reclamation, and he was upset that the Bureau's communications with Upper Basin officials were not similarly restricted. Chapman, worried that the conflict between the Park Service and the Bureau of Reclamation might flare up again, decided the best way to keep the peace was to muffle Drury. In December he approached the Park Service director and offered him a new position in the department. As he later recalled the confrontation, Chapman said, "Newton, you are such an ardent conservationist. I want you to be promoted up to Assistant to the Secretary . . . and become my real advisor on all these conservation matters." Drury, however, was not impressed with the title of "special assistant"—he knew well enough that he wouldn't have anything special to do. In February he tendered his resignation, to become effective on the first of April.[21]

Most conservationists were unhappy both with Drury's resignation and the circumstances which had led up to it. The *New York Times* expressed the opinion of most wilderness advocates when it praised Drury for the "vigor and determination [with which he] fought to preserve the integrity of the national park system," while the *Washington Post* criticized Chapman for the "subterfuge" with which he handled the affair.[22] Some conservationists began to speculate that the Park Service director had been forced out of office for taking too strong a stand against the Echo Park dam. Waldo Leland, chairman of the National Parks Advisory Board, expressed the feelings of many of his colleagues when he stated, "Conservationists throughout the country have every reason to be perplexed and indignant and anxious." Shortly after they heard the news of Drury's resignation, Leland and several other representatives from the advisory board forced a meeting with Chapman. The secretary mollified the group by granting them permission to openly consult with the Park Service on the Dinosaur controversy. The advisory board, in return, agreed not to incite further controversy in the Interior Department.[23]

Not everyone was shedding tears over Drury's departure, however. Most westerners were delighted, because they interpreted the move as evidence that Chapman would not tolerate any opposition to his decision on the Echo Park dam. One western paper guessed that Drury had been fired for supporting DeVoto's widely read "falsification" and surmised "Mr. Drury has been fired because of his power-hungry fact-twisting propagandizing, and that is that." The *Vernal Express* reported happily, "Drury's resignation will speed Echo Dam."[24] A few conservationists were also glad to see Drury go, not least among them former Secretary of Interior Harold Ickes. Ickes was furious with Drury after learning from Chapman about the secret memorandum of understanding he had signed with Page in 1941. Ickes claimed that Drury was "just the opposite" of the crusader elegized by the *New York Times* and characterized him as "one who prefers to say 'yes' on occasions when he should utter a resounding 'no.'" Ickes believed Drury had been "worse than useless" during the Echo Park controversy, but felt hopeful that "there might be a possibility of our securing a revision of the Dinosaur order if we went about it now that Drury is to pass from the scene."[25]

The real circumstances surrounding Drury's resignation remain somewhat unclear. One conservationist believed Ickes may have pressured Chapman into the move because Drury had refused him permission to use a Park Service lodge in Acadia National Park. Chapman later intimated that he had been furious about the secret memorandum and dismissed Drury because the director continually denied having signed it. DeVoto investigated the matter himself without any success, concluding only that it was "an ugly situation" orchestrated in such a way that no one could do anything about it.[26]

But even if the reasons behind it remain unclear, Drury's departure had a significant effect on the Dinosaur controversy. With the Park Service director out of the picture, Utahns seemed sure that the dams would be constructed and lost no time in pressing their advantage further. The editor of the *Vernal Express* boasted, "Lodore Canyon will be a popular resort when Echo Park Dam is constructed," no doubt imagining all the while a flock of tourists pumping money into the local economy.[27]

With the resignation of their chief, the National Park Service seemed just as certain that the dams would be built. In February, one day after news of Drury's resignation hit the papers, the Park Service submitted a plan to develop the recreational resources of the proposed Echo Park and Split Mountain reservoirs. The plan totaled nearly $12.5 million, with the funds earmarked for improvements that were to include, among other things, eighty-five miles of road construction, seven beaches with floating boat docks, and a guest lodge.[28]

Private conservation organizations were not so quick to accept the reservoirs as inevitable. A number of organizations, notably the National Parks Association, the Wilderness Society, and the Izaak Walton League, continued to condemn the Dinosaur dams in their respective membership bulletins. Chapman was undoubtedly eager to prove to his critics that, in spite of his decision on the Echo Park dam, the Interior Department would continue to champion the national park system. On March 7, he issued Order No. 2618, prohibiting all Department of Interior bureaus and agencies from making water development studies in any national park, monument, roadless area, or wildlife refuge without prior consent of the secretary.[29] Chapman also extended the review period on the Bureau of Reclamation's report from the statutorily required ninety days. Judging from a letter he would later write to the secretary of the Vernal Chamber of Commerce, he gave serious consideration to the recommendations that

Within the Canyon of Lodore, a series of rapids—Disaster Falls, Hells Half Mile, Triplet Falls—fills the canyon with the sound of roaring water, thrills boaters, and reminds us of a time now past when rivers ran free. *Photo by Jack Boucher, Dinosaur National Monument.*

came back. "My approval of the report," Chapman asserted, ". . . did not and does not mean that I intend to disregard the views of other agencies and interests and to adopt it without change as my final report, irrespective of adverse opinions. Such an attitude would make a mockery of the review process established by law and by interagency agreement. On the contrary, I take very seriously the objections raised to the proposed report as a result of its review."[30]

The proponents of the dam would not realize for several months just how much weight Chapman was willing to give to these objections. Throughout the spring and summer, the Upper Basin states had returned favorable comments on the development plan and urged immediate authorization of Echo Park dam.[31] Other reports, however, were more critical, including reviews submitted by the U.S. Geological Survey, the Fish and Wildlife Service, and the Department of Agriculture. The most scathing review came from the Army Corps of Engineers, which criticized the project's "unrealistically high" secondary benefits and the extent to which irrigation would be subsidized by power revenues.

Some of Chapman's closest advisors in Interior began to question the plan as well. Foremost among these was Assistant Secretary Dale Doty,

National Park Service Director
Conrad L. Wirth.
Photo by Abbie Rowe,
National Park Service,
Historic Photograph Collection.

who told members of the Sierra Club in November that he was personally opposed to the construction of a dam at Dinosaur. The statement would prompt one booster from Vernal to complain, "Doty . . .
should be sentenced to Dinosaur Monument until he repents," but Doty
remained unshaken. Soon afterwards, he would tell Chapman that "in
no case should the dam be approved."[32] New Park Service Director
Conrad Wirth also pleaded with Chapman to reverse his earlier decision. "Dinosaur is the only area in the national park system that the public can get down into and see and enjoy a grand beautiful canyon such
as this from the river," Wirth wrote in a passionate memo to his boss.
"There is no more beautiful canyon in the country. We must keep it this
way."[33] Chapman himself threw some doubt on the project less than
two weeks after Doty's remarks to the Sierra Club. Speaking before a
meeting of the National Audubon Society in New York, Chapman
assured the conservationists in the audience, "I sincerely hope that we
might work out a solution whereby the Echo Park and Split Mountain
dams need not be built in the monument." To the dismay of the dam's
boosters, Chapman followed up his remarks to the National Audubon
Society by appointing a task force to investigate the conservationists'
claim that alternative sites could spare Dinosaur.[34]

Proponents of the dam reacted bitterly, but carefully directed most
of their anger at the National Park Service. "How the Park Service and
the nature lover groups can influence the powers in Washington to
block the development of water and power . . . is a mystery," opined
the editors of the *Vernal Express*. In an emergency meeting which
immediately followed Chapman's remarks to the Audubon Society, the
Vernal Chamber of Commerce resolved to initiate its own propaganda

campaign in favor of the dam. The editors of the *Vernal Express* offered their assistance and volunteered the services of everyone else in town as well. "If the nature lovers want a good fight," they warned, "there is no group in the world that likes one better than the Westerner." The following summer, Wirth personally visited Vernal to explain his agency's position to a hostile audience of city councilors, members of the Chamber of Commerce, and local business leaders. He left no doubt that he was going to fight the Echo Park dam, and then he beat the quickest trail out of town. Vernalites were "completely dumbfounded" by Wirth's impudence and sent Chapman a telegram declaring that the Park Service should be made to "abide by your decision to build Echo Park." Although Wirth had to take a good deal of the heat for Chapman's apparent change of heart, the dam's proponents were not pleased with the secretary either. B. Frank Ward, secretary of the Vernal Chamber of Commerce, later expressed this displeasure on behalf of the dam's proponents. "We are happy that Mr. Wirth came to Vernal and made the statements that he did," Ward wrote to Chapman, "because he is apparently the first man from the Department of Interior who has the strength of his convictions."[35]

Chapman tried to defend himself from his western critics. To Senator Arthur Watkins he pointed out, "The people of Utah will profit more if their water and power needs can be met without sacrificing the National Monument than they will if we have to sacrifice it." However, his last statements on the Dinosaur dams were notable chiefly for their ambiguity. In a letter to Congress, he recommended authorization of the Glen Canyon, Flaming Gorge, Navajo dams, and "facilities at Echo Park or an alternate site, to serve the purposes intended to be served by the Echo Park unit."[36]

The elections of 1952 gave both sides in the Echo Park controversy reason for renewed hope, since Chapman's opinion of the project would be moot if a Republican was elected president. The conflict in Korea and the emerging Cold War hysteria were undoubtedly the most important issues in the presidential campaign of 1952. Although these issues overshadowed concern about western resource policy, they indirectly impacted the Upper Colorado River Storage Project. A year earlier, Straus took advantage of foreign policy concerns to remind Chapman, "In view of the existing national emergency, consideration should be given to the suitability of the power features of storage projects as they relate to national defense," and he singled out Echo Park dam as one project which would "assist in meeting [power] demands for defense purposes." B. Frank Ward of the Vernal Chamber of Commerce considered opposition to the Echo Park dam to be almost treasonous and wondered "if the subversive forces who are working against this nation don't have a lot to do with it."[37]

For all their ideological differences, Republicans and Democrats arrived at just about the same conclusion on the Upper Colorado River Storage Project during the 1952 campaign. Both parties favored some form of water and power development and differed only on how that development should be administered. The Democratic presidential candidate, Adlai Stevenson, continued Truman's emphasis on public power and federal development of western resources. Exactly where the preservation of wilderness fit into his conservation program was difficult to establish. Because Stevenson was not well versed on western resource policy, Truman took up much of his campaign burden in the western states. While in Colorado, the president tried to pander to conservation sentiment by stating, "I think that the Dinosaur National Monument . . . ought to be preserved. In fact, it ought to be enlarged. After this election we will enlarge it to accommodate the dinosaur wing of the whole Republican Party. We will fill it up with old Republican fossils. Then maybe we could change the name of it to the Republican National Dinosaur Monument. The Republicans are living so far in the past they never in the world will catch up with the progress of this nation of ours." The quip better demonstrates the Democrats' belief that decentralized management of western resources was an outdated philosophy than it does any firm commitment to preservation of undeveloped areas. Truman told the same audience, "If you want the Upper Colorado storage project, you had better vote Democratic."[38]

Republicans tried to attract voters with a message of fiscal responsibility, decentralized bureaucracy, and faith in private enterprise. In the West, these planks in the Republican platform typically translated as a commitment to "return" the public domain to western states. While such a program might jeopardize federal reclamation appropriations, it suggested that the western states would be free to do with their scenery what they wished. At the same time, Republicans promised that water development would continue as a venture between federal and local authorities. Republican presidential nominee Dwight Eisenhower was the most prominent advocate of this sort of "partnership." Under the democratic administration, Eisenhower claimed, the federal government was so deeply involved in water development that it did everything "but come in and wash the dishes for the housewife." Eisenhower told a western audience, "We need river basin development to the highest degree, but not at the expense of accepting super-government in which the people in the region have no voice. . . . We want this to be done through partnership . . . bringing in the federal government not as a boss, not as your dictator, but as a friendly partner, ready to help out and get its long nose out of your business as quickly as that can be accomplished."[39]

Western timber companies, oil and mineral prospectors, and livestock industries, which were all increasingly disgruntled with federal restrictions on the use of public lands, strongly supported the Republican philosophy. So, too, did many advocates of an Upper Colorado River Storage Program. Senator Arthur Watkins of Utah had earlier claimed to be "very happy indeed" that Democrats had authorized the Echo Park dam, but like most western Republicans he was wary of a permanent federal presence in his state. "I do not want to be placed in the position of approving the desire of the Bureau and the Department to own and operate the projects in perpetuity," he warned. "I still believe that after the people have paid for the program by their purchase of power and their purchase of water rights they should be recognized as equitable owners of the project."[40] Despite the obvious benefits of federal reclamation programs, many westerners resented the federal presence that came along with those benefits. They preferred to receive federal reclamation funds without losing local autonomy. DeVoto, for one, had long marveled at this western ambivalence and succinctly captured the prevailing western desire in a single sentence: "Get out and give us more money."[41] But Republicans up for election in 1952, with their faith in private enterprise, were more sympathetic to western pleas for local autonomy in resource development, and they overwhelmingly won national, state, and local offices in the western states.

Conservationists who had been opposing the Echo Park dam must have watched the election returns come in with a tinge of apathy, for neither of the presidential candidates seemed a champion of their cause. On one hand, Truman's Democratic administration had already endorsed the Upper Colorado project, and there was no reason to believe that Stevenson would reverse that decision. On the other hand, conservationists must have been concerned that a Republican emphasis on partnership could too easily turn into a "giveaway" program in which western resources, including land reserved in the park system, would be opened to full exploitation. Preservationists who believed that the "highest use" of some portions of the public domain was to leave them *undeveloped* still seemed to lack a political voice. The fact that both parties favored some sort of development plan spoke volumes about the utilitarian values that shaped American resource policy in the early 1950s.

Eisenhower's victory set the stage for a Republican purge of the Interior Department and a renewed battle over the Upper Colorado Storage Project and the Echo Park dam. But unless the conservationists could bring to bear a different set of values on the controversy— values that recognized undeveloped wilderness as a resource to be protected as well as exploited—the chances of stopping the dam were slim indeed.

CHAPTER FIVE

Voices for the Wilderness

After the election of Dwight Eisenhower, the western states waited expectantly for the president-elect to nominate a new secretary of the Interior. Because of the strong emphasis Eisenhower had given to partnership in the development of western resources, most westerners believed the new secretary should come from their own ranks. Speculation in the weeks immediately following the election centered around Utah Governor J. Bracken Lee and Governor Dan Thornton of Colorado as likely choices.[1] Of course, conservationists fighting the Echo Park dam were not at all eager to see either of these Upper Basin officials take the job.

Eisenhower's eventual selection of Oregon's Governor Douglas McKay came as a surprise to many. The loyal Republican party man was not well known on the national political scene. Newspapers reported that before serving as governor he had owned Doug McKay Chevrolet, a car dealership in Salem. Like most businessmen who enter the public sector, he disdained rampant government spending and budget deficits. Yet no one seemed to know much about his philosophy of resource use. The *Wall Street Journal* reported that McKay was "a proponent of public power and government reclamation projects," but he had never staked a firm position on the proposed Hells Canyon dam on the Snake River, a project which had caused quite a bit of controversy in Oregon during his term as governor.[2] Proponents of the Upper Colorado Storage Project were undoubtedly disappointed that the new secretary was not from the Upper Basin, but they at least were pleased that McKay was a westerner and that he held a businessman's view of resource use. One telling indication of McKay's conservation philosophy came in a statement he made to lumbermen in which he referred to trees as crops. "I have no sympathy for those mistaken souls who preach that our forests should be socialized and turned over . . . for . . . stewardship," McKay claimed. "Neither have I any particular sympathy for the person who screams about free enterprise and the rights of individuals to own forest land and then abuses that privilege or does nothing constructive."[3] McKay's aversion to "stewardship" of federal lands and his emphasis on "constructive" private ownership did not endear him to most conservationists, who quickly became wary of the new secretary. Their apprehension only increased when Oregon Senator Wayne Morse warned them that McKay was "a well-recognized stooge of the tidelands thieves, the private utility gang, and other selfish interests which place materialistic values above human values."[4]

No one doubted that McKay would make sweeping changes at the Department of Interior. Throughout Truman's administration, the Interior Department had been criticized by conservative Republicans as a "seedbed of socialism." One business magazine, speaking for business interests throughout the nation, but especially those in the West,

Secretary of Interior Douglas James McKay. *Photo by Abbie Rowe, National Park Service, Historic Photograph Collection.*

claimed that "the activities and ambitions of this all-embracing department have grown until its planners are well along the road . . . to making all Americans within its domain . . . subservient to government."[5] The communist witch-hunt led by Senator Joseph McCarthy still gripped Washington, and it seemed no Republican could resist attacking a program that had implications of federal planning or collectivism. Even Eisenhower would make a passing reference to the "creeping socialism" dominating the Interior Department. When the president-elect's choice for secretary of Interior was announced, former presidential candidate Thomas Dewey wrote to congratulate McKay. "I am looking forward," Dewey wrote, ". . . to the wonderful job I know you will do in slaying the Socialist dragon of the Interior Department."[6]

To slay that dragon, McKay would have to make a number of key personnel changes in his department. His most important appointment, perhaps, was his selection of Ralph Tudor as undersecretary. Tudor, a graduate of West Point and a former U.S. Army engineer, managed a successful private firm when McKay approached him about the job. Tudor normally eschewed politics, and when he was asked to take the post, he almost deferred. "What," he replied incredulously, "is an under secretary?" When he finally accepted the position, Tudor brought to the Interior Department a savvy business instinct, a strong aversion to inefficiency, and expertise in dealing with the kind of technical information that would support reclamation projects. His connection to West Point would also help mollify criticism leveled at the Bureau of Reclamation by the Army Corps of Engineers. McKay would later place his trusted undersecretary in charge of all intradepartmental correspondence on the Dinosaur issue.[7]

To the relief of conservationists, McKay chose to retain the services of Park Service Director Conrad Wirth, perhaps to avoid the trouble some of them promised to make if Wirth was thrown "to the political wolves."[8] Michael Straus, however, was not as fortunate. The commissioner of Reclamation was probably the man Republicans were most eager to dispose. As strong a proponent of public power as there ever was, Straus had been a thorn in the side of fiscally conservative Republicans, who seethed at his disregard for whether the Bureau's projects were economically feasible. "I don't give a damn whether a project is feasible or not," Straus reportedly told one group of Bureau employees who showed apprehension about a particularly wasteful project. "I'm getting the money out of Congress and you'd damn well better spend it." In 1949, Straus's enemies pinned an obscure provision to a public-works appropriation bill that withheld the commissioner's salary. Much to their chagrin, the independently wealthy Straus remained at his job, without pay. "I have done what no good Republican has been able to do," Straus bragged to William Warne after the incident, "and that is to unite the Republican party on at least one platform and provide them with one program—to wit, who can fire Straus first."[9]

Straus resigned his post before McKay had the pleasure of dispatching him. Before stepping down, however, Straus fired a departing shot at the new Republican administration. The Bureau of Reclamation had prospered, he claimed, "as a bipartisan effort under the Square Deal, the New Deal, and the Fair Deal." Now Straus thought it in danger of being "consigned to the evolving philosophy of the Big Deal."[10] Most conservationists probably would have been happy to see that prediction come true, but the federal reclamation program had far too much support in the West to simply disappear, even under an administration which had pledged to cut back federal spending. Conservationists must have held their breath as rumors spread that George Clyde, Utah's chief water engineer and one of the Upper Colorado Storage Project's strongest proponents, would replace Straus. McKay passed over Clyde, however, in favor of another engineer from the Upper Basin—Coloradoan Wilbur Dexheimer. Unlike Straus, an easterner who had been a professional newspaper man before coming to Reclamation, Dexheimer had spent his entire professional career in the Bureau.[11] The selection of Dexheimer and Tudor suggests that McKay meant to justify his reclamation projects on their technical merits rather than through publicity campaigns or emotional appeals. This seemed a reasonable strategy for dealing with a Republican Congress that had declared its intention to curtail federal spending and had demonstrated an aversion to public power projects—reasonable, that is, as long as the Bureau maintained its reputation of technical competence.

With the major players in the Dinosaur controversy suddenly cleared from the stage, conservationists redoubled their efforts to defeat the Echo Park dam. As the first few months of McKay's term unfolded, however, it became increasingly clear that their access to the new secretary would be more restricted and that their private lobbying efforts would be less effective. To make matters worse, the national publicity campaign against the dam had stalled, and conservationists were finding it difficult to rouse the public sentiment they would need to defeat it. What they needed to find was some way of taking the Dinosaur controversy directly to the people.

Conservationists found what they were looking for in the person of David R. Brower. In December 1952, Brower was appointed executive director of the Sierra Club, one of the oldest and most respected conservation organizations in the country.[12] Although it had played a prominent role in fighting development proposals for a number of national parks in California, by the end of 1952 the Sierra Club still was not a very active participant in the battle to save Dinosaur. Board member Richard Leonard traveled to Dinosaur in the summer of 1950,

David Brower.
Dinosaur National Monument.

and the following summer photographer Philip Hyde made an extended visit to collect photographs that the club could use for publicity purposes, but Echo Park was a long way from the club's headquarters and its core of support. With only about seven thousand members, the club was a relatively small organization. New members could not join without a sponsor, and sponsors were selective enough that one commentator noted wryly, "In those days you didn't apply to join the Sierra Club, you were properly invited."[13] Many club members were more interested in outdoor recreation than politics. When a past president of the club assured members back in San Francisco that Dinosaur was "just canyons and sagebrush," the issue was seemingly settled. Most club members felt that they had little personal investment in the monument and believed there was no reason to involve themselves in the Dinosaur battle.[14]

However, Brower's appointment as the Sierra Club's first professional staff member followed on the heels of a self-conscious renewal of purpose within the club to determine whether it should emphasize outdoor recreation or wilderness preservation. Many club members thought the Sierra was becoming *too* accessible and wanted the club to object more stridently to roads and other intrusions into undeveloped areas. In 1951, the Sierra Club Board of Directors decided to amend the official purpose of the club to reflect the growing concern for wilderness preservation. The club's original statement of purpose—"to explore, enjoy, and render accessible the mountain regions of the Pacific Coast"— was changed to read "to explore, enjoy, and protect the Sierra Nevada and other scenic resources of the United

States." The change of purpose entailed both an emphasis on preservation and an expansion from a regional to a national focus.[15]

Brower, for one, was eager to extend the Sierra Club's sphere of influence outside California. Dr. Harold Bradley, serving at the time as president of the Sierra Club, made the Dinosaur controversy Brower's primary goal. In the summer of 1951, Bradley's son Steve and a friend had floated the Green and Yampa rivers, accompanied by Vernal riverman Bus Hatch. At the site of the Echo Park dam, Bus pointed out ladders coming down the canyon walls, and told Steve of the Bureau of Reclamation's plans to build a dam that would flood the canyons through which they had just traveled.[16] Steve was aghast, and at his urging the entire Bradley family met Bus the next summer for a trip down the Green River. Harold Bradley, seventy-three years old when he stepped aboard Bus Hatch's raft, made a film of the trip which he showed to Sierra Club members in California. The most excited viewer of all was David Brower. "It knocked my hat off," Brower would later say of the film. "I knew I had to get there."[17]

Brower had enough influence with the board of directors to steer the club towards the Dinosaur controversy. "They were willing to see what I could do," Brower later recalled matter-of-factly, "and that was one of the things I thought we ought to do."[18] Before being appointed executive director, Brower had served on the editorial board of its bimonthly *Bulletin* and led numerous High Trips in the Sierra Nevada. He relied heavily on his previous experience as he tackled the challenge of bringing the Dinosaur controversy to a national audience. His first action was to arrange a series of Sierra Club raft trips through Dinosaur modeled on the outings he had led in the past, only on a much grander scale. In instigating these river trips, Brower perhaps recalled John Muir's strategy for protecting the Sierra Reserve. "If every citizen could take a walk through this reserve," Muir had written, "there would be no more trouble about its care; for only in darkness does vandalism flourish." In his unsuccessful campaign to prevent the flooding of Hetch Hetchy, Muir had tried to get as many people to see it as possible. "None as far as I have learned, of all the thousands who have seen the Yosemite Park, is in favor of its destructive water scheme, and the only hope of its promoters seems to be in the darkness that covers it," Muir wrote.[19] Brower similarly believed that the key to saving Dinosaur was to get as many people to see it as possible. Not only would the trips generate publicity, but they would also give lie to the oft-heard claim that Dinosaur was inaccessible and its rivers dangerous.

The first of these Sierra Club river trips occurred in the summer of 1953. Guided by riverman Bus Hatch, more than two hundred Sierra Club members floated the Green and Yampa rivers that summer, including Brower and his two sons, Ken and Bob. On one trip, photographer Charles Eggert captured Dinosaur's scenic grandeur in a color film called *Wilderness River Trail*. At the suggestion of Conrad Wirth and Fred Packard, Eggert had already made a shorter film entitled *This Is Dinosaur*. *Wilderness River Trail* was a more ambitious project. Brower and Eggert spent several weeks collaborating on the script, and a number of conservation organizations distributed it in the following months, introducing hundreds of thousands of people to Dinosaur's canyons. As Brower and Eggert would both later take pride in noting, the dam's proponents credited this film as being the single most effective piece of propaganda used by conservationists in their campaign against the Echo Park dam.[20] The river trips also attracted quite a bit of media attention. Sierra Club member Martin Litton used the trips as the focus for a series of articles for the *Los Angeles Times* and, later, for the *San Francisco Chronicle*. *National Geographic* magazine conducted its own river trip and in 1954 carried an extensive article on Dinosaur, including a slew of color photographs. The article's anti-dam message was implicit rather than explicit, but it was real nevertheless, and it reached a wider audience than any piece of conservationist propaganda had since DeVoto's *Saturday Evening Post* article back in 1950.[21]

Thanks to a "splendid job of diplomacy" on the part of Bus Hatch and Dinosaur Superintendent Jess Lombard, the trips were not restricted to the converted, either.[22] On one trip, *Vernal Express* editor William B. Wallis and several representatives of the Vernal Chamber of Commerce joined the Sierra Club members. After the trip, Wallis wrote an article in which he acknowledged the scenic grandeur of the canyons. Brower was heartened. "I hope I detect in this account a note which presages full appreciation by the Vernal people of something which I, a native Californian, believe to be a more spectacular and beautiful experience than anything we have in California," he wrote to Wallis.[23] In a letter to the secretary of the Vernal Chamber of Commerce, Brower explained, "Leave these canyons the way they are, provide reasonable access to them, schedule a variety of trips on the river, and the world will beat a path to your door." Reminding him of the thousands of dollars Sierra Club members had spent in Vernal since the raft trips began, Brower pointed out that the economic growth created by tourism was just as real as the economic growth Vernal was expecting from the Echo Park dam. "All we are urging you to do is eat your cake and have it," Brower concluded. "At less cost you could get your power and water and your reservoir recreation. And you can still have the unspoiled canyons."[24]

Despite the cheerful diplomacy in Brower's letters, Vernal residents remained staunchly in favor of the dam. Don Hatch, who helped his father conduct the river trips, would later comment, "All we had to do was take someone through [the monument], then comment that a dam might flood it, and they were horror-stricken."[25] Actually, the reaction usually depended on the prejudices which one carried aboard the boat. In 1950, for example, Bus had taken fifteen members of Vernal's Echo Park Dam Committee on a river expedition, after which they concluded that a reservoir would do little damage to the scenic values of the canyons and would enhance the monument by making it more accessible.[26] Bus privately disdained the idea of flooding Echo Park, but because he was a builder and his livelihood depended on winning construction contracts in Vernal and surrounding communities he refrained from speaking out against the dams. He knew the prevailing local attitude was verging on fanatical. His son Don, however, was a student in Salt Lake City. Don had no essential ties to Vernal, so he eagerly took up the anti-dam crusade. According to his father, Don's vocal denunciation of the dam "stirred up quite a hornet's nest" in Vernal. Being an agitator, however, had its price. On more than one occasion, Don was threatened with bodily harm if he set foot in Vernal. One unkind relative wrote to the *Vernal Express* to denounce Don's troublemaking and officially kick him out of the family. Apparently the local fervor for the dam could outstrip the bond of kinship.[27]

If the river trips didn't convince local residents that the Echo Park dam was unnecessary, the publicity they generated certainly *did* reach a national audience that was more receptive to the idea of wilderness preservation. And many of those who experienced Echo Park for themselves added their voices to a protest that was growing louder each day. Yet it would soon become evident that the strategy of protecting wilderness by bringing people into it was problematic. To many people, seclusion was an essential quality of wilderness. By increasing the number of visitors in a wilderness area, it seemed conservationists were in danger of destroying precisely that which they were trying to protect. The Sierra Club's river trips in Dinosaur made the paradox more striking than it had ever been. Joe Penfold captured the essence of the dilemma when he responded to a friend who worried about bringing too many people into Dinosaur. "There is the danger of Echo Park being trampled to death," Penfold admitted, but he concluded, "we would be hypocritical if after all the talking we have done, we tried to keep it for a few wilderness lovers alone."[28]

Brower also showed some inconsistency in his thoughts on this issue. In a letter to Dinosaur Superintendent Jess Lombard, Brower warned that the monument's wilderness qualities could be irreparably damaged by a radio communications tower at a creek known as Jones Hole. The tower would allow park rangers to more effectively patrol the monument and better protect the safety of its visitors.

Instead of acknowledging that the Sierra Club trips may have helped make the tower necessary, Brower wondered how it would look from the raft. "I myself am alarmed at the prospect of what a radio communication station might bring to one of the most beautiful spots on the river," he stated. "In the heart of the wilderness country along the river, I do wonder if electronics and the appurtenant devices are not pushing us a little too far."[29] Brower, like Penfold and most other wilderness advocates, favored "keeping mass recreation developments out of the national parks," even though this position conflicted with his efforts to build a political constituency for wilderness. [30]

The proponents of Echo Park dam exploited this inconsistency in their opponent's thinking by emphasizing that a flatwater reservoir would be accessible to more people than the river rapids. Such a reservoir might "alter" Dinosaur, but it would hardly ruin it. In fact, the dam's proponents even argued that dams would *enhance* the monument. In a pamphlet entitled "Tomorrow's Playground for Millions of Americans," the Upper Colorado River Commission implied that there was no good to a park which nobody visited. "Only by storing and putting to beneficial use the river waters which run through it," the brochure stated, ". . . can Echo Park and surrounding country truly become a park."[31] Secretary McKay seemed to hold the same view. Early in 1954, McKay declared that two of the three national park areas with the greatest number of visitors in 1953 "both benefited greatly from federal dam developments." Even Conrad Wirth emphasized the need for tourist facilities in the parks through his so-called "Mission 66," a ten-year program to make the parks more accessible.[32]

Conservationists in private organizations generally disdained the Department of Interior's lack of appreciation for the "deeper ideology" of wilderness preservation. One conservationist complained that officials in that department regarded the national parks as "glorified tourist resorts, playgrounds for the titillation and entertainment of the public."[33] Howard Zahniser later would assert that "the very nature of our preservation effort in the National Park System is threatened" by developments intended to make the parks more accessible.[34] To conservationists like Zahniser and Brower, accessibility was not the primary value to be advanced. "Whether or not all of us use these places is not important," Brower insisted. "In Salt Lake City they don't junk the Temple because so many more people enter the Tabernacle." More important was to keep wild areas in their natural state. The Echo Park dam might make Dinosaur accessible to more people, but in doing so it would ruin the wilderness qualities that made it worth preserving. The dam would not "enhance" or merely "alter" the monument—it would destroy it. Accepting "the amazing statement that Echo Park dam will not destroy Dinosaur, but will only alter

Dinosaur," Brower would later claim, was essentially the same as accepting the statement that "cutting the 3,000-year-old Big Trees and making them into grapesticks" would only alter Sequoia National Park. "Maybe 'alter' isn't the right word," Brower concluded. "Maybe we should just come out with it and say 'cut the heart out.'"[35]

When McKay announced that he would earmark $21 million for recreational facilities at Dinosaur once the dams were constructed, he either displayed a poor grasp of the ideology behind wilderness preservation or he shrewdly exploited the ambivalence most conservationists felt toward physical "improvements" in the national parks. Everyone admitted that Dinosaur was not accessible. The few roads that existed in the monument were made impassable by inclement weather. A rancher who lived near Echo Park, not known for his timidity, said that when the road got rough, he'd "hesitate to fly a kite over it."[36] Yet most conservationists were outraged at the news. "Is this the new policy regarding our national parks?" asked Arthur Carhart. "First you wreck them, then you spend ten times as much on amusement park type recreation installations?" In disgust, he compared McKay to "a man who has dressed his wife in flour sack clothes all her life, then promises her on her deathbed that he will buy a satin-lined casket and dress her corpse in silks."[37]

In the three years since Oscar Chapman's hearings on the Echo Park dam, conservation organizations had formed a coalition which demonstrated a remarkable degree of cohesion. In July 1953, however, it appeared that this solidarity might be cracking. Bestor Robinson, a director of the Sierra Club and a member of special advisory panel to the Secretary of Interior, sent a formal letter to Assistant Secretary of Interior John Marr suggesting that the Echo Park project be abandoned but that the Split Mountain dam be constructed as planned. Although he felt strongly about preserving Dinosaur, Robinson admitted to Marr that, "If I were a resident of Vernal, Utah, I must frankly concede that I would develop a different viewpoint on the problem." Robinson proposed the compromise because he believed conservationists could not prevent construction of the dams and that their efforts should therefore be focused on saving the most scenic portions of the monument. "The Upper Colorado Basin states are lined up solidly behind the Colorado Storage Project as now presented," Robinson pointed out to his friend Richard Leonard. "This block of votes is sufficient to secure approval of even uneconomic or undesirable construction projects." Although he may have felt that he was merely being realistic, Robinson knew that many of his colleagues did not feel the same way. "I am certain of only one thing," Robinson confided to Leonard, almost apologetically, "and that is that I will be thoroughly condemned by most of my conservation friends."[38]

Conservationists were shocked at Robinson's letter and moved quickly to prevent his proposal from becoming a political liability in their crusade to save Dinosaur. Brower worried that "there is serious danger that the letter will be seized upon and exploited by those who oppose Dinosaur National Monument as evidence that conservationists are split." He warned, "Although Bestor Robinson has assumed 'sole personal responsibility' for the letter in question . . . it is not possible for an individual who has been elevated to a position of leadership in conservation . . . to do other than speak ostensibly for conservationists, no matter how detailed the disclaimers." On August 16, the executive committee of the Sierra Club unanimously resolved to defend the principle of park inviolability. "After full review of alternate proposals and suggestions of compromise," the committee declared, "construction of *both* these dams must be opposed." Richard Leonard wrote to Marr in August to emphasize that the Sierra Club was united on this point. Brower, in the meantime, sent a general memorandum to "cooperators" to urge them to maintain their solidarity.[39]

Robinson's compromise proposal was never seriously considered by either side of the Echo Park debate. The Split Mountain dam was intended mainly to be a re-regulating dam. It would generate some base-load electricity, but its main purpose was to capture peak releases from the Echo Park reservoir. Without the Echo Park dam, there was no reason to build the dam at Split Mountain. Like the rest of his "conservationist friends," Robinson clearly wanted to prevent any dams in Dinosaur, if possible. On its surface, Robinson's compromise proposal seemed to represent merely a difference in strategy. On closer examination, however, it seems that the proposal contains the seeds of a philosophical difference that would divide conservationists in the final stages of the Echo Park controversy. For most conservationists, the *principle* of park inviolability had to be protected at all costs. For a significant minority, however, including Robinson, the greater concern was saving scenic resources, regardless of whether the scenery was specifically set aside and protected by law. Robinson, seemingly, was willing to sacrifice the principle of inviolability in order to save the "better portion—qualitatively and quantitatively" of the monument.[40] Although conservationists preserved what was rapidly becoming an effective political coalition, before long they would become deeply divided on how to best use its power.

If conservationists were initially apprehensive about the resource policies that McKay and the Republican Congress would adopt, it wasn't long before their apprehension turned to outrage. Early in 1953, the Republican Congress transferred offshore oil lands from the federal government to the states. In March, Congressmen Frank Barrett of Wyoming and Wesley D'Ewart of Montana introduced bills to transfer jurisdiction of public grazing lands to states and private users, a move that would have eliminated federal management of millions of acres of public land.[41] To conservationists, it seemed as if the Republican policy of partnership had already turned into a land grab. DeVoto, who had used his *Easy Chair* column to frustrate a similar western land grab in 1948, was disturbed to see Barrett and company back in the saddle. Barrett had "carried the ball for the land grabbers" for several years, but DeVoto was confident that conservationists could rally enough opposition to "keep the bastids [*sic*] off balance." DeVoto perhaps did not expect the criticism that Barrett's proposal drew from many westerners, who believed the bills went so far that they would jeopardize consideration of more important legislation, including the Upper Colorado Storage Project.[42]

By the end of 1953, conservationists had dubbed the secretary of Interior with a new nickname: "Giveaway McKay." Indeed, McKay's later comments would seem to justify the label. "A vigorous effort will be made to advance and enlarge the surveying and classification of the public lands," McKay told one western audience early in 1954, "in order to determine whether they are available for entry."[43] However, most officials at the Department of Interior believed that they had been unfairly branded with the giveaway label. High-ranking officials such as Ralph Tudor believed that the real land grabs were conducted by federal agencies which locked away the resources of the public domain. "If there was ever a 'giveaway' program in existence in this country," Tudor complained to his diary, "it has been during the last twenty years when the regime here in Washington has been 'giving away' the assets of the country for the benefit of a few."[44]

McKay's western supporters were prone to agree. Proponents of the Upper Colorado Storage Project were certainly not enamored of federal reservations which limited resource development. They considered Echo Park to be the biggest "grab" of all. As early as 1946, the *Vernal Express* decried the "government grabbing of public lands" and warned that "the past federal action of acquiring twice as much land as was required for Dinosaur National Monument may set a pattern for the future unless checks are placed on the bureaus who set the grabs in motion."[45]

Although conservationists were perhaps justifiably afraid that transfer of the public lands would lead to over-exploitation and misuse of natural resources, McKay's first year in office was probably not as dreadful as his critics contended. The secretary's willingness to transfer federal land to the states, for instance, was in all ways consistent with the Taylor Grazing Act and case law governing management of the public domain. Moreover, McKay risked alienating his western supporters when he called the Barrett grazing bills "lousy." He told

Wesley D'Ewart that the administration would "do something for the cattleman, but not at the expense of the rest of the people." He wanted D'Ewart to understand that "there's to be no giveaway program of public land."[46] Unfortunately, McKay did not make this statement publicly; nor did he draw much attention to his decision not to authorize Glacier View Dam in Montana's Glacier National Park or to his steadfast refusal to grant logging rights in Olympic National Park. Whether or not the criticism was justified, however, conservationists clearly considered themselves at odds with the administration's resource policies by the end of 1953.

In December, the Mid-Century Conference on Natural Resources convened in Washington, D.C. President-elect Eisenhower had given the mid-century conference a strong endorsement in 1952, but by the end of 1953 the political climate had turned so bitter that Presidential Assistant Sherman Adams convinced Eisenhower to withdraw White House sponsorship.[47] During the conference, McKay tried to pacify conservationists by telling members of the National Parks Advisory Board that he was undecided on the Dinosaur dams. He conveniently failed to mention that he had already approved a report drafted by Undersecretary Ralph Tudor which claimed that the best combination of alternative dam sites would result in a net evaporation loss of 100,000- to 200,000-acre-feet per year. On December 10, just a few days after his statement to the advisory board, McKay submitted the report to President Eisenhower, saying that the Department of Interior endorsed the Upper Colorado Storage Project in its entirety. "If conflicting interests did not exist, I would prefer to see the Monument remain in its natural state," McKay stated. "However, I do feel that if the dam is built, the beauty of the park will be by no means destroyed and it will remain an area of great attraction to many people."[48]

Conservationists were alarmed by the decision and angry with McKay. One even suggested that they try to impeach McKay for "dereliction of duty."[49] The secretary knew his approval of Echo Park dam would further alienate him from conservationists, but he tried to persuade them that his decision would not endanger the park system. "By no stretch of the imagination can our recommendation on Echo Park be construed as a precedent for the invasion of other existing parks," McKay asserted. "No one in this administration is going to build a dam in Yosemite, in Yellowstone, or run a pipeline into Crater Lake."[50] Conservationists, however, took little comfort in McKay's assurances. Three separate bills had already been introduced in Congress to authorize the initial phase of the Upper Basin Storage Project, and in January the House Committee on Interior and Insular Affairs was scheduled to hold formal hearings on the project. The Interior Department's favorable recommendation increased the likelihood that Congress would approve the project.

Conservationists, afraid that their three-year campaign against the Echo Park dam might have been too little too late, anxiously geared up for the coming legislative battle. Howard Zahniser was concerned enough to cut a photograph from the Wilderness Society's membership bulletin just before it went to press. He used the space to scrawl a hasty note to its readers, urging them to write to President Eisenhower and protest the Echo Park dam. In an article entitled "A Call to Battle," Fred Packard predicted "a struggle that may determine whether the conservation and nature protection gains of the past half century are going to be held, or whether a coterie of selfish men will deprive the American people of their most valuable possessions."[51] Arthur Carhart noted that Interior's recommendation would "precipitate the controversy" and declared ominously, "unquestionably this is the show down."[52]

CHAPTER SIX

"No Holds Are Barred"

In January 1954, the House Committee on Interior and Insular Affairs convened hearings on the Upper Colorado River Storage Project (CRSP). The project was sweeping in scope. It included half a dozen main-stem storage dams and a dozen participating projects at an anticipated cost of over one-and-a-half billion dollars. Yet for nearly four years it had been debated and delayed because of opposition to a single dam. Throughout that period, proponents of the project resolutely clung to their belief that the Echo Park dam was the "wheelhorse" of the entire plan and refused to consider alternative sites. National conservation organizations had been just as adamant in their unwillingness to compromise the national park system, and they had shown an unprecedented degree of cooperation in their opposition to the dam. By 1954, however, the project had a lot of momentum behind it. Two separate Interior administrations—one Democratic and the other Republican—had given it their blessing, and President Eisenhower also seemed to like the plan. Congressional approval was its final hurdle.

As the Echo Park debate shifted from an administrative to a legislative arena, the stakes mounted. By this time, the struggle transcended the question of whether a single dam should be built—it encompassed two conflicting views of resource development. The Bureau of Reclamation and the Upper Basin states believed that western resources should be developed for local benefit. Conservationists, in contrast, believed that the highest use of some public land was to leave it undeveloped for the benefit of the nation as a whole. The Echo Park dam had become a symbol in an ideological battle to determine the very

Colorado Congressman Wayne Aspinall.
Dinosaur National Monument.

definition of "conservation." George Pughe, a member of the Colorado Water Conservation Board, urged people in the Upper Basin states to keep fighting for the Echo Park and Split Mountain dams. "I think we should get our toe in the door with the two big dams," Pughe told them. "If we back down now on Echo Park, we'll just get shoved around some more on other things." Congressman Wayne Aspinall of Colorado later voiced the same sentiments when he warned reclamationists that if they agreed to delete Echo Park dam from the Upper Basin Storage Project, they would be "handing conservationists a tool they'll use for the next hundred years." Opponents of the dam also believed that the outcome of the Echo Park debate would set a precedent of lasting value. They had, after all, involved themselves in the controversy to prevent a precedent of developing the national parks. As the congressional debate heated up, remarks such as those made by Pughe and Aspinall steeled their resolve. "The dam proponents are out to smash the conservationists this time and steam-roller them—give them a stinging drubbing to be remembered," Arthur Carhart would later declare in an attempt to rally his colleagues. "No holds are barred."[1]

Through the end of 1953, the Izaak Walton League, the Wilderness Society, and the National Parks Association had been leading the campaign against Echo Park dam. Under the leadership of its new executive director, however, the Sierra Club was becoming an increasingly active participant in that fight. Brower himself had taken a vigorous interest in the Dinosaur controversy after his first river trip through the monument, and by the beginning of 1954 he was undoubtedly one of the leaders in the fight against the Echo Park and Split Mountain dams. It was his testimony before the House Subcommittee on Irrigation and Reclamation in January 1954, however, that would propel Brower to the very forefront of the Echo Park controversy.

Brower spent the first several days of the hearings as an observer, listening critically to engineers from the Bureau of Reclamation and to officials from the Upper Basin states make their case for the Upper Colorado River Storage Project.[2] The Bureau's engineers made it a point to extol the virtues of the Echo Park dam, with a delegation of seventy Utahns on hand to lend political weight to their technical testimony. Their justification for the Echo Park dam had not changed substantially. The engineers admitted that it would be *technologically* feasible to build a storage dam somewhere else, but they emphasized that alternatives were wasteful. The members of the subcommittee, most of them westerners, knew already that water was the most precious resource in the arid Colorado basin, and that inefficient use of water would limit the long-term development of the region. They were therefore likely to agree that evaporation losses were the "most important single factor" to consider when comparing Echo Park to any alternative sites. According to the Bureau's data, even the best combination of alternative sites would result in net evaporation losses of some 165,000-acre-feet annually. Undersecretary Ralph Tudor tried to impress upon the congressmen the practical necessity of the Echo Park dam. "I share the concern of those who would preserve the beauties of Dinosaur National Monument in their present natural state, but . . . I believe the conservation of the water in the interest of the nation is of the greatest importance." Bureau engineers also mentioned the power benefits of the Echo Park site, but they left no doubt that evaporation was their primary concern. "In the final analysis," Tudor stated, "the increased losses of water by evaporation from the alternative sites is the fundamental issue upon which the Department has felt it necessary to give any consideration to the Echo Park dam and reservoir."[3]

After listening carefully to two-and-a-half days of testimony in support of the project, conservationists got a chance to make their case against the dam. For the most part, their arguments were also old ones. Brower was the striking exception. Like his colleagues, Brower began by noting the importance of keeping the national parks inviolate. The principle of wilderness preservation, he claimed, was "one of the finest steps in land-use administration ever devised in the history of the world." Untrammeled wilderness, Brower argued, should be protected even if everyone didn't appreciate its scenic value. "There are probably quite a few," Brower admitted, "who would not care to . . . see any of the wild back country of the parks, who would not care to climb in a rubber boat and float down the rapid and calm stretches of Dinosaur's beautiful canyons. But to some people," he noted, "this very trip has been the finest scenic experience they ever had." After describing Dinosaur's scenic values, however, Brower shifted gears dramatically. The testimony to the subcommittee, he complained, "has consisted in large part of a single Bureau looking upon its own work and pronouncing it good." Brower wanted to turn a critical eye on the technical data the Bureau was using to support the CRSP. Although General U.S. Grant III had already proposed a series of alternative dam sites, and others had questioned some of the economic aspects of the project, no one had ever challenged the validity of Bureau of Reclamation's own technical data. Brower was about to. "In spite of all the study to date," Brower told the subcommittee, "the project is not yet shaken down."[4]

Calculating the rate of evaporation from a standing body of water is a complicated process, involving many scientific uncertainties and complex mathematics. Although Bureau of Reclamation's methods for calculating evaporation rates were admittedly imperfect, its engineers were among the most qualified people in the world to make those kinds of estimates. The Bureau had recently collaborated with the U.S. Geological Survey and the Navy Department on the most advanced report on evaporation ever published. Brower, on the other hand, had dropped out of the University of California without finishing his sophomore year. In August 1953, when he first turned his attention to the Bureau's evaporation figures for the Upper Colorado Storage Project, Brower asked Luna Leopold whether the Bureau could have made a miscalculation. Leopold, the son of noted wilderness advocate Aldo Leopold and a respected hydrologist working at the U.S. Geological Service, warned Brower that it would be the "height of folly" to try to challenge the Bureau's engineers on evaporation figures, because the Bureau was "much more likely to be right than any consultant that the Sierra Club could ever muster."

"If the Sierra Club gets into the problem of suggesting alternative sites for the Echo Park and Split Mountain Dam you are going to let yourself wide open," Leopold warned. He believed that conservationists could successfully oppose the Echo Park dam solely on the basis that it would flood a national monument. "It would be my suggestion that the Sierra Club forget about the question of alternatives and put some more steam into the basic issue of encroachment," he concluded. "That is a good argument and the Bureau people know it."[5]

Brower ignored the advice. Before coming to Washington, he had consulted with his brother and his father, who were both engineers, and with his friend Walter Huber, a former official of the Sierra Club and president of the American Society of Civil Engineers. He also scrutinized the Bureau's report from cover to cover with an editor's eye for mistakes. "The House document was about this thick," Brower later recalled, holding his fingers several inches apart. "I don't know where my copy is, but I had more notes in the margins; again and again they were lousing things up."[6] Now, sitting before the congressional sub-committee, Brower was understandably nervous. "I didn't take my pulse that day," he later recollected, "but I knew it was there." Citing the Bureau's own base figures, he insisted that the alternative sites did not evaporate an extra 165,000-acre-feet, as Undersecretary Ralph Tudor had testified, but actually *reduced* evaporation losses by 20,000-acre-feet.[7] When Brower finished his testimony, the members of the subcommittee stared back at him in disbelief. The Bureau of Reclamation attracted some of the most talented engineers in the country. To question their technical expertise seemed absurd. Brower had hardly begun his critique of the Bureau's evaporation rates, however, before the hearing abruptly adjourned for the day, as if the members of the subcommittee were uncomfortable or unprepared to handle testimony on technical data.[8]

When the hearing reconvened the next morning, the Bureau of Reclamation arranged to have a blackboard on which Brower could display the evaporation figures. The Bureau also arranged to have Cecil B. Jacobson, the chief engineer of the Upper Colorado Storage Project, look over his shoulder as he did so. Considering the difficult circumstances, Brower handled himself with aplomb. "My point," he stated, "is to demonstrate to this committee that they would be making a great mistake to rely upon the figures presented by the Bureau of Reclamation when they cannot add, subtract, multiply or divide. I am not trying to sound smart," he added, "but it is an important thing." The subcommittee members were skeptical of Brower's challenge. Wayne Aspinall of Colorado asked incredulously, "You are a layman and you are making that charge against the Bureau of Reclamation?" Other subcommittee members challenged Brower's credentials as well. Despite his early and somewhat coy disclaimer that he was just "a man who has gone through the ninth grade and learned his arithmetic," three different congressmen asked Brower whether he was an engineer. At one point, the chairman of the subcommittee warned Brower not "to impugn the ability of the engineers." Congressman William Dawson of Utah was aghast. "If Mr. Tudor is such a poor engineer as you seem to claim he is, I'm surprised that he ever got that . . . bridge down in your town to meet in the center."[9]

After Brower worked through his figures, Jacobson commented at length on his testimony, assuring the subcommittee members that "engineering is not just ninth grade arithmetic." Before he was done, however, Jacobson had to concede that the Bureau's figures contained an error and a misprint. Instead of the 165,000-acre-foot loss that Tudor had claimed, Jacobson admitted the loss should be only 70,000-acre-feet. Brower was happy to have made his point. "I am glad that is corrected now," he said. "Would it have been corrected had I not raised the question? And would it not be well to get competent engineers to raise the questions about a lot of these other facts?" But Brower also made it clear that he opposed the Echo Park dam regardless of the evaporation rate. "The last impression I am leaving you with is one of figures," Brower told the subcommittee, but he added, "I did not come here to do arithmetic. I came here to try to advocate a principle." His closing remarks were pointed. "Before we sell out our parks," Brower asked, "shouldn't we attack real waste first? Wasteful irrigation methods, for one thing. Wasteful pollution for another. Wasteful soil erosion due to small-watershed mismanagement. The list of wasteful things we do is nothing to be proud of. When we've whittled that list down, then—and not until then—let's see where else to pare." Echo Park, Brower declared, belonged as much to future generations as to the people who wanted to flood it. "Let's leave to them the choice of selling their birthright," he pleaded. "They won't even have a chance to choose unless we leave them that birthright, unless we bring about an enlightened approach to the parks in this, their darkest hour."[10]

Not surprisingly, Brower came under scathing attack from proponents of the CRSP. The Upper Colorado River Commission, for instance, ridiculed the "utter stupidity of attempting to calculate evaporation losses by the use of ninth grade arithmetic." If the dam's boosters were concerned about Brower's testimony, none of them admitted it publicly. "Everything looks very, very good from here," a delegate from Vernal declared when the hearings drew to a close. "In fact, it looks better than I have ever seen it."[11]

Despite the general optimism in the Upper Basin states, however, the evaporation issue became a major source of embarrassment for the Department of Interior. In 1950, the Bureau of Reclamation had been claiming that evaporation losses would exceed 350,000-acre-feet for the best combination of alternatives to the Echo Park dam. In the spring of 1951 regional engineer E. O. Larson questioned the reliability of that estimate in a memo to his superiors in Washington. Their reply made it clear that the Bureau was on shaky ground. "We are not confused, as you state, about evaporation," they told Larson, "but frankly we are confused by the incomplete, vague, or contradictory reports from your office." Larson decided to reduce the evaporation figure to a more conservative 300,000-acre-feet, but in 1953 the chief engineer in Denver questioned that figure as well, stating that it could not "be sustained in the event of a critical review . . . by an unprejudiced agency." By January 1954, the Bureau had pared the figure to between one- and two-hundred-thousand-acre-feet, with Undersecretary Ralph Tudor giving 165,000-acre-feet as the department's best estimate. While Tudor admitted that "there may be some error due to a shortage of experimental data," he assured McKay that "the calculations are reasonable."

After Brower's testimony, Tudor must have stayed awake nights wondering how he could most gracefully dislodge his foot from his mouth. On March 9, he wrote to the House subcommittee to inform its members that the additional evaporation resulting from the best series of alternatives should be adjusted down to 70,000-acre-feet. Only a few weeks later, after further review of the evaporation figures, Tudor had to *again* write to the subcommittee, putting losses now at only 25,000-acre-feet. At that point, members of the subcommittee may have harkened back to the question Brower's raised during his testimony: "How much lower can they go—*and still be wrong?*" Knowing that the Bureau of Reclamation's credibility was severely compromised, Tudor ordered it to recalculate all of its evaporation figures.[12]

Although they did not question the competence of Bureau engineers as brazenly as Brower had, other conservationists did not let the evaporation issue pass by unnoticed. Howard Zahniser praised Tudor for "one of the most courageous public statements I have noted in the years I have been in Washington" and diplomatically acknowledged the Department of Interior's great "prestige and reputation." But he also pointed out that Interior's endorsement of the project was based on the erroneous figures, and he urged Tudor to "not leave the scenes until you have been able to work out a correction of the damage that has been done."[13] Although the evaporation issue provided a convenient excuse to jettison the Echo Park dam and move forward with the rest of the CRSP, the Department of Interior refused to reverse its recommendation. "For reasons which were given in my testimony and by other witnesses before your subcommittee," Tudor told its chairman, "we do not consider . . . a substitute justifiable."[14]

Although the evaporation issue did not change the Interior Department's position on the Echo Park dam, it did dispel one of the Bureau of Reclamation's most important justifications for the project. More importantly, it cast doubt on the Bureau's technical expertise. Even the staunchly pro-dam *Salt Lake Tribune* could not resist a lighthearted jab at the Bureau when it awarded its regional director a rubber slide rule for "stretching the truth" about the evaporation figures.[15] Some conservationists felt that they could now talk about alternative development schemes and have them taken seriously. But not everyone would agree that that was a sound strategy.

Brower's testimony before the House subcommittee slowed the momentum behind the Echo Park dam, but it hardly eliminated it. In February, Senator Arthur Watkins and three other members of the Utah congressional delegation called on President Eisenhower to impress upon him the importance of the Colorado River Storage Project. The meeting lasted forty minutes behind closed doors. The Utahns emerged smiling, but kept the details to themselves. When one reporter asked Watkins to disclose the problems he discussed with the president, the senator insisted they had mentioned only "state matters." Then with a smile he added, "We have no problems in Utah." Watkins's jovial mood made it plain that he was satisfied with Eisenhower's commitment to the project. The *Vernal Express* declared that the delegation had produced the "additional blessings of the administration" for the Central Utah Project and the Echo Park dam. In March, Budget Director Joseph Dodge trimmed some of the more uneconomical elements from the plan that McKay had approved, reducing the projected cost by almost a half billion dollars, but he did not touch the Echo Park dam. A few days later, President Eisenhower officially threw his weight behind the project as well.[16]

Eisenhower's endorsement of the CRSP undoubtedly influenced congressmen who had not yet made up their minds about the project. The chairman of the House Interior Committee had already predicted his committee would report favorably on the CRSP. In May, only one week after Undersecretary Tudor officially acknowledged the erroneous evaporation figures, the prediction came true; by a thirteen to twelve vote, the House Interior Committee approved the Upper Colorado Storage Project, including the Echo Park dam.[17]

Brower was incensed. Looking to vent a little frustration, he wrote to Presidential Assistant Sherman Adams, urging the administration to withdraw support of the objectionable dam. "Mundane men will not know how to take and enforce that stand," Brower challenged. "It will not please little men. It requires grasp, imagination, vision, and statesmanship."[18] The implication Brower was brashly suggesting, of course, was that the administration's continued approval of the project would be proof that both Adams and the president were "little men." Such statements surely did not win the conservationists any friends in the White House.

In June, the Senate Interior Committee prepared for its own hearings on the Upper Colorado Storage Project. When the clerk for that committee informed Fred Packard of the National Parks Association that the hearings were to be brief and that only three witnesses would be allowed to testify against the project, Packard insisted that Brower be one of them. By this time, the Sierra Club executive director was undoubtedly leading the fight against the Dinosaur dams. But Packard also advised Brower that the tone of the hearings would in all likelihood be polite and warned him that "it is quite probable that if any of us raises questions that sound critical of the Bureau or others, it will damage our case."[19] Indeed, Packard's advice suggests that some conservationists were starting to worry that Brower's uncompromising, confrontational tactics might damage their political credibility.

Still, the conservationists were making headway. Before the Senate hearings began, Packard confided in Brower, "I simply cannot see how they have the strength to get [an authorization bill] through Congress this session."[20] The fact that the House Interior Committee had already approved an authorization bill did not worry him, since westerners generally dominated the Interior committees of both chambers of Congress. Conservationists knew the project would have much tougher going on the floor of the House if and when it ever emerged from the Rules Committee, and that neither chamber of Congress was likely to pass an authorization bill before the Bureau of Reclamation finished recalculating its evaporation figures.

Conservationists intensified their lobbying campaign throughout the spring and summer of 1954. Howard Zahniser, executive secretary of the Wilderness Society, orchestrated the effort. "He was one of the two people who put glue in the conservation movement," Brower later remembered. "He and Ira Gabrielson got more people to work together who otherwise would have gone off in different directions."[21] Zahniser's previous experience as a feature writer and as director of an information bureau in the Department of Agriculture enhanced his ability to communicate the conservation message. Many of Zahniser's colleagues also brought impressive public relation skills to the Dinosaur campaign. William Voigt of the Izaak Walton League, for instance, had formerly worked as a bureau chief for the Associated Press and served as public relations manager of the Carnegie-Illinois Steel Company. Although it is difficult to say for sure how much influence the conservationists wielded on Capitol Hill, their language was undoubtedly more persuasive than the technical idiom of the Bureau's engineers.

The lobbying efforts hit a potential snag when in June a U.S. Supreme Court decision made it illegal for organizations which received tax-deductible donations to use that money to influence Congress. Most conservation organizations fell into that category, so the opponents of Echo Park dam established several new organizations—including Trustees for Conservation, the Citizens Committee on Natural Resources, and the Council of Conservationists—solely to continue lobbying.[22] Meanwhile, conservationists continued their publicity campaign against the dam, with Brower leading the way. In February, Brower had collaborated on a special edition of the Sierra Club *Bulletin* dedicated exclusively to the Echo Park controversy. Brower then followed up the Sierra Club's earlier film with another entitled *Two Yosemites*, comparing Echo Park to the unsightly reservoir in Yosemite's Hetch Hetchy valley. "The parallel with Dinosaur was so beautiful that we worked on that constantly," Brower later said of

Hetch Hetchy. The Sierra Club continued its river trips through Dinosaur as well, allowing more than nine hundred people to float down the Green and Yampa rivers that summer.[23] As a crowning touch, Brower began looking for contributors for a book about Dinosaur. *This Is Dinosaur: Echo Park Country and Its Magic Rivers,* which came out the following year, celebrated Dinosaur in photographs and in a series of essays edited by Wallace Stegner. A copy of the book was sent to every member of Congress, along with a hard-hitting brochure put in the book at the bindery. Brower called it "probably the most potent brochure I ever put together."[24]

In July, the Senate Interior Committee passed an authorization bill by a vote of eleven to one, as expected. The bill was briefly debated on the Senate floor, but no vote was taken. The House bill, meanwhile, was still tied up in the Rules Committee. When the 83rd Congress adjourned in August, the Bureau of Reclamation had not yet finished recalculating its evaporation figures, and the authorization bills died a quiet death. Conservationists could be pleased with their work. "We've forged a team for ourselves that has worked together better than any other . . . in conservation history," Brower declared at an annual convention of Associated Sportsmen of California at the end of the summer. "The signals aren't perfect," he admitted, "we can still stand a little more drill on fundamentals, and we'll need to add more plays to our repertoire; but it's already an effective national team."[25] Mail to key congressmen had been running heavily against the dam throughout the spring and summer. One Republican was moved by this outpouring to remind Presidential Assistant Sherman Adams of the enormous sentiment in favor of preserving the national parks. "It is worth remembering," he warned, "that the combined population of Utah and Colorado . . . is barely equal to the number of people who visit Yellowstone and Yosemite every year."[26] A national constituency for the national parks had been aroused.

Proponents of the Echo Park dam tried to stem the huge tide of public opinion against the dam with a publicity campaign of their own. In February 1954, the Vernal Kiwanis Club started a letter writing campaign in support of the Echo Park dam through which several thousand letters reached key members of the House and Senate Interior committees. A Utahn by the name of J. H. Ratliff had already made a film of the "dangerous" rapids of the Green and Yampa rivers to "combat propaganda by such people as Bernard DeVoto who have never seen the canyons."[27] The Upper Colorado River Commission distributed its own film, entitled *Birth of a Basin.* The commission also fostered the creation of the Aqualantes, a group of "water vigilantes pledged to back the Colorado River Storage Project." In exchange for one dollar bills to fund the pro-dam publicity campaign, the

Aqualantes handed out lapel buttons shaped like a sheriff's star. With state headquarters in Salt Lake City, Grand Junction, Albuquerque, and Cheyenne, the Aqualantes were clearly a regional organization. Much of their publicity, however, was aimed at a national audience. *"All* citizens of the United States stand to benefit from this great project," the group insisted. "It is a tremendous factor in national defense. It will aid the national economy. It will provide water and power for the growing West. Its benefits are literally countless."[28]

The Aqualantes handed out thousands of sheriff stars in their first few months of existence and inspired a flurry of political action in the Upper Basin states. By the end of 1954, grassroots groups from all four of the Upper Basin states announced that they would each contribute $10,000 dollars to the Upper Colorado River Commission's publicity fund. The commission was delighted when the Navajo nation kicked in an additional $10,000 of its own.[29] The economic plight of the Navajos was a strong card in the pro-dam publicity campaign. The CRSP was expected to irrigate about 125,000 acres of land on the Navajo reservation in New Mexico. The chairman of the Navajo Tribal Council, envisioning hundreds of new irrigated farms for his people, stated, "This [project] will enable us to support ourselves with the dignity and human satisfaction to which every citizen is entitled." The Aqualantes were quick to pick up on the argument. "This project will help the Indians help themselves," said one pamphlet entitled "Straight Talk." If assistance for the Indians was not enough, the pamphlet also noted that "in the long run, [the project] will save the government money because it helps the Navajos to become self-supporting, instead of having to be supported by government expenditures."[30] As they benevolently advanced the cause of the Navajo people, the Aqualantes were careful not to overlook the interests of eastern taxpayers who would eventually subsidize the CRSP.

By the end of 1954, the Upper Colorado River Commission was working toward a fund of $200,000 for "educational purposes" and had changed the name of the Echo Park dam to the "Yampa-Lodore Unit" to eliminate connotations of a pastoral garden on the edge of the river.[31] Even the pro-dam propaganda, however, was influenced by the swelling tide of conservation sentiment. Very little of it appealed to the traditional pioneer justification of subduing nature.[32] Instead of condemning wilderness, most proponents of the Dinosaur dams tried to appeal to the same conservation sentiment that was being used against them. The Echo Park dam, they insisted, was a *true* conservation project, and anyone who said differently was misleading the public. "Many of us who have been a part of the conservation movement in the West," one Utahn claimed, "are at a loss to understand the motives of conservationists opposing a project which will result in such a material gain to conservation objectives and principles." A pamphlet put out by the Upper Colorado River Grassroots, Inc., noted that the Western Association of State Game and Fish Commissioners favored the dam because "the post-project wildlife and recreation values . . . will be far greater than the undeveloped river now possesses." The same pamphlet quoted G. E. Untermann, a former ranger-naturalist at Dinosaur serving at the time as director of Utah's Field House of Natural History. Speaking for himself and other conservationists from Utah, Untermann claimed, "Our lives have been devoted to conservation, and we see the need for the proposed project. We know the area and realize that its beauty won't be destroyed."[33]

However slanted this grassroots publicity campaign may have been, congressional leaders from the Upper Basin states eagerly threw their weight behind it. Utah Senator Arthur Watkins told an audience in Salt Lake City that "despite the obvious organized opposition" to Echo Park dam, Congress was going to pass an authorization bill. Apparently he didn't care that mail to both the Interior committees was overwhelmingly opposed to the dam; he rationalized that "one singular factor in the controversy is that the farther the letter writer lives, the more vehement his protest becomes."[34] Watkins and his western colleagues in Congress were not about to allow easterners to impose their highbrow philosophy of resource management onto their constituents. In December, Watkins traveled to Washington to receive congratulations from President Eisenhower on his handling of the McCarthy censure committee. He used the opportunity to make a strong appeal for the Upper Colorado Storage Project and requested that the president mention it in his coming State of the Union Address.[35] A few weeks later, Eisenhower came through with the endorsement.

At the beginning of 1955, a new Democratic Congress geared up to consider yet another round of authorization bills. After five years of delay, most proponents of the Upper Colorado Storage Project were confident that Congress would finally take substantive action on an authorization bill. Conservationists, too, could sense a climax approaching. One way or another, Congress would resolve the Echo Park controversy before the year was over.

CHAPTER SEVEN

Strange Bedfellows

On January 18, 1955, Clinton Anderson of New Mexico, chairman of the Senate Interior Committee, introduced yet another authorization bill for the Upper Colorado River Storage Project. The bill was identical to the one which his committee had overwhelmingly approved a year earlier, but this time it was co-sponsored by ten other senators from the Upper and Lower basins. By February, four similar authorization bills had been introduced in House.[1] All of them included Echo Park dam, and all of them were opposed by national conservation organizations.

Despite this strong show of unity on the part of the Upper Basin congressional delegations, the dam's proponents found it increasingly difficult to maintain their political cohesion as the legislative debate dragged on. Although westerners were united behind the principle of federal reclamation, they all wanted to prevent their neighbors from staking too big a claim on the limited water that was available. Their coalition entailed a variety of interests, some of which conflicted, and there had always been the danger that some members would become preoccupied with their own pet projects to the point of indifference toward other aspects of the project.

California presented the most serious obstacle to the Upper Colorado River Storage Project. As the state with the most irrigated acreage, California was a strong proponent of federal reclamation in general. But California was adamantly opposed to the CRSP because upstream diversions would seriously reduce the amount of water it could expect from the Colorado River. The project would also increase the concentrations of salts and alkalis in the water that California did receive, and it would possibly lead to water shortages in Lake Mead and reduce the power capacity of Hoover Dam. Acting to protect its own interests, California threw its weight against the Upper Colorado Storage Project. By 1954, California lobbyists had already spent over a million dollars trying to defeat it.[2]

Proponents of the Upper Colorado project attempted to win California over to their side by pointing out that upstream dams would greatly reduce the amount of silt deposited into Lake Mead and greatly extend the active life of Hoover Dam. At the same time, they accused conservationists in general—and the Sierra Club in particular—of being puppets in California's bid to monopolize the Colorado River. DeVoto, who was much amused by this line of argument, facetiously told Arthur Carhart that California was paying him precisely $122,614.52 for his services.[3] However much conservationists resented being characterized as California's patsies, they didn't hesitate to pursue a convenient alliance with California. Brower met several times with Northcott Ely, the leading attorney for California's Colorado River Board, to compare data on the Upper Colorado Storage Project. "I'm not sure whether they approached us or I approached them," Brower later admitted. "We certainly saw that we were together. . . . We would join anyone who could help us save Dinosaur."[4]

In addition to California's opposition, proponents of the Upper Colorado River Storage Project also had to contend with differences of opinion among the Upper Basin officials. When the governors of the Upper Basin states met in Cheyenne, Wyoming, to discuss the project in January 1955, Colorado Governor Ed Johnson expressed some concern about the manner in which the benefits of the project would be divided. He pointed out that 72 percent of the water in the Colorado River system originates in Colorado, and he told his colleagues that he wanted to capture water close to its source with several small dams in order to increase irrigation benefits for his own state. Johnson also proposed dividing the power revenues from the storage dams based on the formula used to apportion water among the Upper Basin states and using those revenues to pay for each state's participating projects. The proposal caused an exchange of barbed language between Johnson and Utah Governor J. Bracken Lee, who declared that the proposal was an "argument along the same line given to me by California interests in many instances." Conservationists noted gleefully that under the terms of Johnson's proposal, the Central Utah Project would be "knocked in the head" since its irrigation features as planned would require up to 70 percent of all power revenues for at least the first twenty-five years of operation, while under Johnson's proposal Utah would receive only 23 percent of the power revenues.

To satisfy Johnson, the Upper Basin governors finally agreed to add twenty provisional projects to the authorization bill at a cost of about $275 million. Johnson dropped his proposal to divide the power revenues, and the issue was seemingly resolved. But Joe Penfold of the Izaak Walton League noted that Johnson's proposal "could very well serve to kill the whole project as presently planned" by bringing to light "very graphically some of the weird economic theories proposed by the Bureau of Reclamation."[5]

Indeed, as the legislative debate heated up, the economic aspects of the Upper Colorado Storage Project were scrutinized more and more closely. Eastern congressmen, in particular, were concerned about the cost of farm subsidies and surpluses at a time when the Department of Agriculture estimated that the nation was swamped with $7 billion worth of food surpluses, which it warehoused at a cost of some $700,000 per day.[6] Conservative Republicans were similarly concerned with the growing federal deficit. Eisenhower had promised to be more frugal than the New Dealers who had run Interior for so long, and his administration had indeed slashed the Bureau of Reclamation's budget to $156 million—less than half of what it had been only five years earlier. Still, most eastern congressmen were not satisfied. Congressman John P. Saylor of Pennsylvania wrote to McKay to express a displeasure he shared with many of his colleagues. "I am sorry to notify you that as far as I am concerned, I can find no change whatsoever in the Bureau of Reclamation," Saylor stated. "The projects presented today are as fantastic, if not more so, as those presented by the Democrats. . . . Dexheimer is continuing the strategy of Mr. Strauss [sic]."[7]

There was no shortage of critics who agreed with Saylor. Free-market advocates had for a long time been insisting that the Bureau of Reclamation shouldered private enterprise out of the water development business, played favorites when it came time to dole out benefits, and poured capital into projects that were not economically efficient. These arguments had fallen into disuse during the long string of Democratic presidents, but after Eisenhower's election they started to make a comeback. One of the most scathing critics of the Bureau's policies was disillusioned New Dealer Raymond Moley, who frequently used his column in *Newsweek* to attack federal reclamation programs.[8] Elmer T. Peterson's book *Big Dam Foolishness,* published in 1954, presented another blow to the Bureau. Peterson focused his attention on the Missouri Valley Project, but in the process he attacked the wastefulness of all large-scale federal reclamation projects. The book so impressed President Eisenhower that he discussed it with freshmen congressman at a luncheon early in 1955.[9] By the time Congress began to seriously consider authorization bills for the CRSP, the *New York Times* had editorialized that its economics made as much sense as growing "bananas on Pike's Peak" and critics such as Bernard DeVoto were insisting that "the entire concept of reclamation needs a thorough overhauling."[10]

There were even critics of federal reclamation from the Upper Basin states. Perhaps the most prominent of these was former Wyoming Governor Leslie A. Miller. As chairman of the Hoover Commission's Task Group on Reclamation and Water Supply, Miller's opinion commanded respect both from government officials and the general public. As early as 1949, Miller had warned American families that they would each be obligated to pay some $1,500 in taxes for federal water projects that were under construction or in the planning stages. He wasted no opportunity to disparage the economic aspects of the Upper Colorado Storage Project. "[It] will eventuate the loss of hundreds of millions of dollars to the general economy of the nation," Miller told the House Interior Committee in March 1955, and he made much the same point in a widely read *Reader's Digest* article a few months later.[11] On the grassroots level, there was a group called the Angrilantes, based in Glenwood Springs, Colorado. Set up in opposition to the Aqualantes, the Angrilantes defined themselves as "taxpayers angry at fool dams by dam fools." Characterizing reclamationists as "pork barrel pirates whose snouts are always in the public trough," the Angrilantes declared they would support "no more dams until our mildewed mountain of food shrinks to a molehill." The Angrilantes certainly did not represent the typical voter in the Upper Basin, and they did not have much political or financial clout. One letter put out by this group noted that with an operating budget of $200 they were going to fight a regional organization that had raised a publicity fund of $160,000. "We can lick 'em," the letter assured its readers, "but we ain't got the $200 yet." The fact that such a group existed at all, however, underscores how pervasive was the discontent with uneconomical, federally financed reclamation projects.[12]

If the economic aspects of most federal reclamation projects were questionable, the economics of the Upper Colorado River Storage Project were particularly dreadful. The cost of delivering water through the Central Utah Project, for instance, would be $793 per acre of irrigated land, with the water user paying only $74—less than 10 percent—of that cost.[13] The difference would be borne by the taxpayer. Other participating projects were even more expensive.[14] Of all the storage dams included in the project, the Echo Park dam had the highest cost per unit of water stored, at $27 per acre-foot of storage. In contrast, the proposed Glen Canyon dam would provide an acre-foot of storage for only $16.20, and other dams would do the same for as little as $9.80. Even the power aspects of the plan were questionable. The power plant in the base of the Echo Park dam, for instance, would generate electricity at a cost of $883 per kilowatt, whereas coal-powered steam plants in Denver and Salt Lake City were already producing comparable amounts of electricity at a cost of only $168 and $166 per kilowatt, respectively.[15]

To make matters worse, the Bureau of Reclamation had a long-running record for underestimating the cost of its projects. The Missouri Valley Project, then under construction, was a dramatic example. Its projected cost had nearly quadrupled after its authorization.[16] In addition, many people questioned the cost-benefit analyses that the Bureau had been using to justify its projects. As early as 1950, the Izaak Walton League had declared, "It is no secret that the bookkeeping at which justification [for most federal water projects] was arrived sometimes has been questionable." That organization's opinion was supported by Budget Director Joseph Dodge, who told Secretary McKay that "the procedures used to compute the secondary benefits of the participating projects . . . would appear to require a fundamental re-examination." By the time Congress turned its attention to the Upper Colorado River Storage Project, most informed critics of federal reclamation would have agreed with Miller's opinion that the cost-benefit analyses which the Bureau used to justify the project were "of little or no validity."[17]

Much of the political opposition to the Upper Colorado River Storage Project was based on the project's questionable economics. Even officials in Utah, such as fiscally conservative Governor J. Bracken Lee, had to acknowledge the problem. "I think your soundest argument is against the cost, and certainly isn't because it is part of the National Park System," Lee told one conservationist. "At least where people live right with that park and prefer the water, I think you have to concede we have some right."[18]

With so many political allies, however, conservationists did not have to concede anything. By uniting with California water interests and fiscally conservative Republicans, the conservationists could conceivably defeat the entire CRSP. Both of these groups had tangled with conservation organizations at one time or another. But most conservationists could overlook those differences for the time being; to save Echo Park, they would have to take up with such strange bedfellows.

As more attention was focused on the economic aspects of the Upper Colorado Storage Project, conservationists found themselves faced with a troubling dilemma. Throughout the Echo Park controversy, they had been insisting that they were opposed only to intrusions in the national park system and not to sound development of water resources. Howard Zahniser, for example, was incensed with Secretary McKay for making a speech in Salt Lake City in which he said conservationists were "opposed to any development which would disturb one tree or displace one mountain boulder."

"How can you so characterize us who have made every effort to make clear that our opposition to your proposals has been only to those two dams which have been proposed for construction within Dinosaur National Monument?" Zahniser asked the secretary angrily. To the House subcommittee he asserted, "Again and again we conservationists . . . have sought to emphasize that we do not object to dams, or to reclamation, or to water storage or hydro-electric power production, but to the use of a particular site, or sites, in the Dinosaur National Monument."[19] Fred Packard of the National Parks Association expressed the same view in a memorandum to members of Congress in which he stated, "Conservation organizations have no desire to impede orderly development of the water resources of the Colorado River, and favor a soundly planned Upper Colorado River storage program."[20] Zahniser, Packard, and other conservationists who argued in favor of sound water development knew that they could not maintain their political credibility unless they sounded reasonable.

However, as the Upper Colorado River Storage Project was held up to closer scrutiny, it began to look less and less like a program of "sound development," and some conservationists were tempted to try to scuttle the entire project. As early 1950, many of these same conservationists had demonstrated some concern about the economic aspects of the project, but at that time they were hesitant to make economics the focal point of their argument. "While we should not consider as determinative the economic aspect," a Sierra Club official stated in 1950, "I believe that it should be considered in drawing that balance sheet."[21] When J. Bracken Lee suggested to Brower in the spring of 1954 that conservationists were on "sounder ground" opposing the Echo Park dam on an economic basis rather than on the basis that it would "injure some park," Brower disagreed. "The Sierra Club has been operating now for sixty-two years," he told the Utah governor, "and [it is] making headway trying to persuade practical men that practicality isn't everything."[22]

A year later, Brower seemed to have had a change of heart. "The principle of park preservation should be able to stand alone, and the day when it can do so may be near at hand," Brower wrote in a Sierra Club booklet that was prepared in the spring of 1955. "But we have been persuaded by practical men that one way to prevent park invasion is to offer alternatives to that invasion." By the summer of 1955 so much criticism had been directed at the economic and technical aspects of the CRSP that Brower and a few other conservationists thought it made sense to emphasize alternative development plans. The most attractive of these alternative plans included a higher dam at Glen Canyon. Brower promoted the high Glen Canyon dam as a means of having sound water development while keeping the parks unimpaired. His logic was that "since they were going to build that dam as the big money-maker, allegedly, in the whole project, then they might as well make it a little higher while they were at it."[23]

Others did not agree with Brower's strategy. When Joe Penfold of the Izaak Walton League suggested that conservationists endorse an alternative Upper Colorado Storage Project in order to stay consistent with their original support for water development, Fred Smith argued against him. "The philosophy is sound enough," conceded the executive secretary of the Council of Conservationists, "but there is no reason to believe that we are in a position to create an acceptable project. Even if we could, we would be most vulnerable, because we are not engineers, economists, nor judges of the best possible use of the water in the Colorado River." John Baker, president of the National Audubon Society, praised Howard Zahniser for avoiding an economic critique of the project in his testimony before the House subcommittee. "This is the most appealing statement on the subject that I have read," Baker told Zahniser. "By contrast, many of the other hearing statements on the same subject bored me to read, because they go on more or less endlessly, arguing about the engineering details, or the advantages of alternate sites. . . . Based on such experiences as I have had with appearances at Congressional hearings, Congressmen also get bored listening to the same old stuff that they have read or heard before, but are tremendously pleased when they are approached as though they were intelligent human beings with an interest in the preservation of beauty and other things worthwhile. In my judgment, you are much more apt to get what you want by such an approach than by the argumentative or statistical one."[24] Other leading conservationists, particularly those affiliated with the Wilderness Society and the National Parks Association, agreed with Baker. Olaus J. Murie, president of the Wilderness Society, went so far as to claim that evaporation figures and other technical data "should be considered irrelevant."[25]

If conservationists paused during the summer of 1955 to assess the progress they had made in the Echo Park battle, they may have been stung by the irony of their situation. Conservationists from around the country had rallied around Echo Park, had set aside their divergent agendas to work together, and, in their unity, had found political power. But now that they had it, they could not agree on how to use it.

In April, the Senate Interior Committee reported favorably on the Upper Colorado River Storage Project, just as it had done the year before. When the bill came to the Senate floor, however, newly elected Democratic Senator Richard Neuberger of Oregon moved for an amendment to delete the Echo Park dam. Neuberger's proposal received surprisingly strong support. Even some proponents of federal reclamation favored the measure, either because they considered the

CRSP project too badly needed to be stalled any further or because they believed that the Republican administration was overdeveloping the Colorado at the expense of other river basins. Thirty senators voted to remove the Echo Park dam from the bill. Fifty-two voted to keep it. With the Echo Park dam intact, the bill came to the floor of the Senate for a full vote. In an action that one reporter called "a log-rolling campaign worthy of a Paul Bunyon," the Senate authorized the Upper Colorado River Storage Project by a vote of fifty-eight to twenty-three.[26]

Despite the Senate's approval of the CRSP, conservationists remained optimistic. "This is a better vote than any of us anticipated," said a newsletter distributed by the Council of Conservationists, implying that their lobbying efforts were indeed having a positive effect. "We are unquestionably gaining ground," Fred Smith assured his colleagues in May. "I have at least a dozen letters from Congressmen who have changed their positions in the last two weeks on the basis of exorbitant costs and immature planning." Conservationists knew that the House of Representatives was more reluctant to spend money on reclamation projects, and therefore believed they would have a much better chance of defeating the project when it came up for vote on the House floor.[27]

However, as they kept their eyes and ears on the treatment of the authorization bill in the House, conservationists began to worry that political bargaining there might again lead to a vote for authorization. Rumors circulated that California's Claire Engle, now chairman of the House Interior Committee, was considering throwing his support to the Upper Basin project out of fear that senators from the Upper Basin states would vote against a reclamation project in Engle's own home district. Similarly, Oregon's Senators Neuberger and Wayne Morse were concerned about the fate of a proposed project in Hells Canyon. In short, conservationists worried that the almost limitless potential for deal-making might prove deadly, because the favor-trading encompassed "almost all of the states west of Kansas . . . where water is a priceless commodity and a magic political word."[28]

To make matters worse, conservationists feared that the dam could be restored in House-Senate conference committee even if the House passed the bill without the Echo Park dam. The only way to be sure the dam's proponents wouldn't attempt that type of end-run was to defeat the entire project. A few conservationists advocated that strategy, even though it meant reversing their earlier position on water development. "You will hear of maneuvers calculated to make it seem as though Echo Park Dam is dead," warned an open letter distributed by the Council of Conservationists. "It will never be dead unless the whole poorly-planned, uneconomical Upper Colorado River Project is voted down by the House of Representatives." Arthur Carhart urged conservationists "to make sure for sure" that the dam was defeated and "to lick the whole Upper Colorado Project on the floor of the House."[29]

Conservationists may have tried to do exactly that if in June, on a motion by Colorado Congressman Wayne Aspinall, the House Interior Committee had not decided to delete the Echo Park dam from the authorization bill. "We hated to lose it," admitted Representative William Dawson of Utah, "but the opposition from conservation organizations has been such to convince us . . . that authorization legislation could not be passed unless this dam was taken out." Then, with the controversial dam no longer an issue, the House Interior Committee approved the CRSP by a vote of twenty to six. Even with the Echo Park dam taken out of the project, however, opposition to the authorization bill was so fierce that Aspinall decided to delay its arrival on the House floor. Dawson, still optimistic, stated, "We are going to get the bill out [of committee] and it will substantially be intact except for Echo Park Dam, which now is in a category where we feel it isn't lost by any means."[30]

When the first session of the 84th Congress adjourned, the House had taken no further action, and it was apparent to everyone that support for the Upper Colorado River Storage Project was fading fast. If conservationists kept up their opposition, they could almost certainly kill any chance it had of being authorized. On the first of November, the governors and Congressional delegations of the Upper Basin states convened in Denver with representatives of the Upper Colorado River Commission to discuss strategies for breaking the legislative bottleneck created by the Echo Park dam. The Upper Basin states now feared that the whole project would be jeopardized if the impasse continued, and at least some of the delegates must have arrived at the meeting open to the idea of burying the Echo Park dam permanently. The idea gained all the additional strength it would need when the delegates saw the previous day's edition of the *Denver Post*. The Council of Conservationists had taken out a full-page ad in the form of an open letter to the Upper Basin Strategy Committee. After enumerating the arguments against the Echo Park dam, the letter stated that conservationists would work to scuttle the entire Upper Colorado River Storage Project if the Echo Park dam, or any secret hopes for it, was not permanently removed. The threat was tempered with reassurance that conservationists "are not anti-reclamationist, and are not fighting the principle of water use in the West," thereby setting the stage for a compromise in which Echo Park dam could be sacrificed in exchange for conservationists' support of the rest of the project, however unjustifiable it may have been on other grounds. On the second day of the meeting, the delegates "resolved that, in the hope of getting action on an Upper Colorado River storage project bill in the present Congress, the Senators and Representatives agree that they will not try to reinsert the Echo Park dam."[31]

Because the CRSP still faced strong opposition from California, from many eastern taxpayers, and from fiscal conservatives, the fate of the project seemed to hinge on whether conservationists withdrew their opposition. Californians tried desperately to convince the conservationists to keep up their opposition to the project. "The conservationists have walked into a trap," California Congressman Craig Hosmer warned in a press release. "By withdrawing their opposition to the bill, the conservationists are permitting themselves to be deceived. They are being lulled to sleep, and they will wake up some morning to find that Echo Park Dam has been built in Dinosaur National Monument."[32] Senator Clinton Anderson of New Mexico wrote to Fred Smith to reassure him that senators from the Upper Basin states would not try to reinsert the dam when the authorization bill was adjusted in conference, and warned him that "the State of California, which hopes to get all the water of the Colorado River for itself, will probably continue to finance a lively campaign against the Upper Basin States." With as much politeness as he could muster, Anderson added, "I would hope that the Council of Conservationists would not lend themselves to that movement."[33]

Conservationists no doubt relished any opportunity to prove that they were not merely hapless tools in California's attempt to monopolize the Colorado River. With a compromise on the table which would save Echo Park, conservationists threw their support behind the Upper Colorado River Storage Project. In return, the sponsors of the authorization bill removed the Echo Park dam and in its place substituted a provision stating that "it is the intention of Congress that no dam or reservoir constructed under the authorization of this Act shall be within any national park or monument."[34]

With the compromise language crafted, Congress was finally in a position to authorize the Upper Colorado Storage Project. "The Sierra Club gave up," David Brower later recalled, "and the opposition to the whole project thereupon collapsed. The other groups—people interested in taxes and interested in keeping water in Southern California—the other organizations just faded away. The Sierra Club was the keystone in that, and the keystone was pulled out and the arch collapsed." His assessment of the situation may give too much credit to his own organization, but it otherwise seems sound. Without the aid of the conservation organizations, opponents of the project could not marshal enough votes to stop it.

On March 2, 1956, as a slightly apprehensive David Brower watched from the press gallery, the House authorized the project by a vote of 256 to 136. Three weeks later a House-Senate conference committee tied up the remaining loose ends. The final version of the bill included four main storage dams—Curecanti Dam on the Gunnison River, Flaming Gorge Dam on the Green, Navajo Dam on the San Juan, and Glen Canyon Dam on the Colorado—and eleven irrigation projects. On April 11, President Eisenhower signed the Upper Colorado River Storage Project into law.

It had taken more than five years, but conservationists had successfully derailed the Bureau of Reclamation's plan to inundate the canyons of Dinosaur National Monument. Forty years after an unsuccessful campaign to save Yosemite's Hetch Hetchy Valley, conservationists proved they had the political muscle to preserve Echo Park. In the process, they persuaded policymakers that the intangible benefits of wilderness preservation could outweigh the benefits of relentless water development. Like most political outcomes, however, the bargain which saved Echo Park was built less on strict adherence to a principle than on negotiation and compromise. Conservationists had not *intended* to strike any bargains. They had steadfastly refused to consider building a dam in another part of Dinosaur; they had not been tempted to redraw the monument's boundaries. From the outset, they had been determined to defend the principles of the national park system.

But like the hero in a Sophoclean tragedy, conservationists met destiny on the road they took to avoid it. To protect the principle that the park system was inviolable, conservationists endorsed a water development plan which was economically and ecologically suspect and which inundated some spectacular scenery, including Glen Canyon. Before long, what had seemed to be the conservation movement's greatest victory would be considered by many to be its greatest sin.

CHAPTER EIGHT

The Place No One Knew

O n October 15, 1956, a group of engineers from the Bureau of
Reclamation gathered near the Utah-Arizona border and, standing on the rim of Glen Canyon, they looked down on the silty ribbon
of water below them with a collective gleam in their eyes. The spot
was not normally the type of place where news was made, being separated from the nearest town of Kanab, Utah, by seventy-five miles
of roadless, rugged, arid terrain. On this particular morning, however,
the Bureau engineers were joined by a handful of newspaper reporters
and radio broadcasters from Salt Lake City and Phoenix. Bureau technicians would soon set off explosives that they had packed into the
canyon's west wall, and the reporters had come to witness what
promised to be a dramatic explosion. Even more importantly, they had
come on behalf of the people of the Utah and Arizona to show their
approval for the construction of a monumental plug in Glen Canyon,
a dam which would finally capture and put to active use the water of
the upper Colorado River.

Meanwhile, more than two thousand miles away in Washington,
President Eisenhower stood in the White House, basking gratuitously
in the spotlight which the Upper Colorado River Storage Project had
cast on his administration. When all was ready, the president
depressed a ceremonial telegraph key and sent a signal to Kanab,
which was then relayed by radio to the site at Glen Canyon. Upon
receiving the signal, a technician there leaned onto a plunger and
detonated the explosives. With a deep rumble that reverberated for
miles up and down the river, huge chunks of Navajo sandstone

sloughed off the canyon wall and came crashing down in a cloud of dust. Six months after Eisenhower had signed the Upper Colorado River Storage Project into law, the construction of Glen Canyon Dam was underway.[1]

David Brower did not pay much attention to the media spectacle in Glen Canyon that October morning. With the Echo Park and Split Mountain dams successfully removed from the Upper Colorado River Storage Project, he had returned to San Francisco and the daily activities of the Sierra Club, which included, among other things, planning mountain High Trips into the Sierra, the Tetons, and the North Cascades. Like most other conservationists, however, Brower kept a wary eye on Dinosaur National Monument.

Despite their pledge to abandon the Echo Park dam, many officials from the Upper Basin states hoped that the dam might someday be built, and a few openly conceded that it had been "only temporarily forsaken."[2] Residents of Vernal were not overly dismayed by what they saw as a temporary setback. When they received news that the CRSP had been authorized, the city erupted in a spontaneous celebration. A procession of cars, ambulances, and a fire truck paraded up and down Main Street with sirens wailing jubilantly. Vernal's mayor told local reporters, "We are all elated at the news . . . even though we don't know as yet all it will do for us. This is part of what we have been waiting for for thirty years. Of course, originally we wanted Echo Park included, and now—I think eventually we will get it."[3]

Dinosaur Superintendent Jess Lombard, who was in Vernal at the time, did not like what he heard. He was concerned enough to fire off a letter to Brower in San Francisco. "I mingled with groups later that evening," Lombard reported, "and the universal theme was 'We have got the Upper Colorado River Storage Project authorized—now we go after Echo Park.'"[4] In the weeks ahead, it became clear that Lombard had good reason to be concerned. Still stinging from their political defeat at the hands of conservationists, Upper Basin officials adamantly opposed two bills introduced in Congress to grant Dinosaur the status of a bona fide national park.[5] Utah Senator Arthur Watkins led the opposition, complaining bitterly that the bills demonstrated "a disturbing breach of faith on the part of conservationists." Ira Gabrielson considered such statements "most disappointing" and ominously implied to the senator that "the attitude and action of the general public" would support the conservation organizations if he renewed the fight for the Echo Park dam. Brower, by now well aware of the political strength of these organizations, also warned Watkins not to raise their ire. "We hope that you will not jeopardize the appropriations necessary to get your historic project underway," Brower stated dryly in a telegram to the Utah senator, "by allowing Congress and the nation's citizens to infer that Echo Park Dam is still so much alive in your mind that you would oppose creation of a great Dinosaur national park."[6]

Watkins knew that Brower was not bluffing, and in all likelihood he took the warning seriously, but it didn't much matter. That fall Watkins lost his seat in the Senate to Frank Moss, a Democrat who promised to renew the fight for Echo Park dam. Utahns knew that Watkins had been the dam's most ardent—and most effective— booster, but they knew equally as well that his hands were now tied by the compromise he had made with conservationists. "The senior senator caused this state to lose Echo Park," Moss told voters, and they responded by dispatching the venerable Watkins. After his election, Moss and other Upper Basin officials—particularly those from Utah— continued to resist attempts to reclassify Dinosaur as a national park. Their opposition was driven by a common motive, and most didn't bother to disguise it. Moss, for instance, responded to one constituent who urged the reclassification by stating flatly, "I too would like to see Dinosaur National Monument become a national park, provided we are able to build Echo Park Dam. Only in this way will Utah receive the water and power to which it is entitled, and the dam and reservoir will enhance the scenic value of Dinosaur." Utah Governor George Clyde brought cheers from the crowd gathered to celebrate the Vernal Unit of the CUP when he told them emphatically, "Don't you think the Echo Park Project is dead—it isn't. It *will* be built."[7]

Utah kept up its vigorous opposition to granting Dinosaur park status until Senator Gordon Allott of Colorado introduced a new version of the bill in 1957. Allott's proposal contained an ambiguous provision on water development, which stated, "Nothing contained in this Act shall preclude the Secretary of the Interior from investigating, under the authority vested in him by the Federal reclamation laws, the suitability of reservoir and canal sites within Dinosaur National Park."[8] Even though similar provisions in President Roosevelt's 1938 proclamation were the root of much of the earlier controversy, many conservationists were tempted to support Allott's bill. Brower had claimed earlier that to have a bill for Dinosaur National Park originate in Colorado "would require a miracle in politics." Wayne Aspinall of Colorado had sponsored one such bill in 1956, but when the measure drew unexpectedly strong opposition from his own state as well as from Utah, Aspinall decided not to renew his sponsorship the next year. Many conservationists wanted to take advantage of Allott's proposal while they had the chance. Horace Albright told one friend that "if we want a Dinosaur park bill, we will probably have to take this one with all its faults, as we have had to do in the cases of such parks as Grand Canyon." Fred Smith, executive secretary of the Council of Conservationists, pointed out that national monuments could be altered or abolished by presidential proclamation and warned his colleagues, "We will be living in a fool's paradise if we think the Upper Colorado bill provides any permanent protection to Dinosaur National Monument."[9]

Other conservationists were less enthusiastic about the Allott bill. Brower reminded his colleagues that "the recent great effort for Dinosaur forged for the conservationists a sharp-edged tool they shall need to wield again." He warned that "the Allott bill dulls that edge badly." Fred Packard, speaking on behalf of the National Parks Association, said the "objectionable phraseology could be interpreted as an expression of intent reversing the earlier one." Packard subsequently came under scathing personal criticism by Smith. Cooler heads eventually prevailed when the president of the Wilderness Society implored fellow conservationists to "keep our arguments on a high level," but the impressive solidarity that conservationists had demonstrated on the CRSP was at long last showing signs of strain.[10]

In September 1957, the Sierra Club officially joined twenty-two other organizations already opposed to the Allott bill, and the campaign for a Dinosaur National Park was abandoned. Allott later confided to Pennsylvania Congressman John Saylor that the language in his bill was intended to ensure that the Echo Park dam would someday be built.[11] That dream faded slowly in Utah. At the end of 1958 the Utah delegation to the Upper Colorado River Commission sought support for a new authorization bill for the Echo Park dam, but the commission was concerned with implementing the rest of the CRSP and refused to stir up any more controversy. Eventually even the *Salt Lake Tribune* came out against the dam in an editorial entitled "Don't Muddy the Upper Colorado Waters." The Vernal Chamber of Commerce, searching desperately for allies, turned in November 1959 for support from the National Reclamation Association. They, too, refused to support anything so controversial. Although the dream faded slowly in Utah, by 1959 it was clear to just about everyone that the Echo Park dam was no longer feasible politically.

With the Echo Park dam no longer a major concern, and with no obvious banner around which to rally, conservationists turned their attention inward and began to consider how they could best advance their cause in the years ahead. Encouraged by the Dinosaur victory, many conservationists wanted to introduce bold, new conservation initiatives. The most sweeping of these new initiatives was Howard Zahniser's proposal to establish a system of national wilderness preserves. Zahniser had put his idea for a wilderness system on hold during the Dinosaur controversy, but once the CRSP was authorized he immediately resumed work on it. "Let us try to be done with a wilderness preservation program made up of a sequence of overlapping emergencies, threats, and defense campaigns," Zahniser told his colleagues. He knew they could never keep the pace.

Conservationists rallied once again behind this proposal, for they realized that unless there was a strong national program to protect wilderness they would exhaust themselves trying to fight each and every development scheme that popped up. When the wilderness bill was introduced in the Senate in 1957, conservationists expected westerners to holler the old arguments about "locking away" resources, and they were not disappointed. But the Forest Service and the Park Service, agencies with which conservationists had worked very closely in the past, also opposed the idea with an unexpected vigor. The wilderness bill put these agencies in an uncomfortable position because it implied that they were not adequately managing their respective jurisdictions.

Some conservationists were in fact frustrated with what seemed to be a misguided conservation philosophy in these agencies. If advancing the conservation cause meant criticizing government officials or their policies, then many conservationists were willing to do just that. Other conservationists, however, were more interested in building bridges than in erecting barricades. The Wilderness Society's Olaus Murie spoke for colleagues who emphasized the need for cooperation and constructive dialogue when he adamantly denounced "saber-slashing military methods" of persuasion.[12]

While conservationists debated strategy among themselves, the rapport they had established with the Park Service during the Echo Park controversy quickly soured. If conservation organizations had served as the voice of the Park Service in 1950, by the end of the decade they were more like its conscience, prodding the agency down paths it might not take on its own. The trouble was that embattled Park Service officials didn't want to hear it. Less than a decade after Newton Drury had asked conservationists to speak for his agency, Park Service officials wanted to silence many conservationists. Their number one pest was David Brower. Horace Albright and Park Service Director Conrad Wirth were so incensed at the scathing criticism Brower directed at the Park Service that Albright tried to convince the Sierra Club Board of Directors to fire their outspoken executive director. The club kept Brower, but in 1959 they passed a resolution that "no statement should be used that expressly, impliedly, or by reasonable inference criticizes the motives, integrity, or competence of an official or bureau." Brower derisively referred to the resolution as a "gag rule."[13]

Brower had always been willing to push the limits of protocol. But his aversion to diplomatic niceties and political patronage was heightened dramatically when he traveled through Glen Canyon for the first time in the summer of 1957. Brower had not been entirely unaware of the area's reputation as the living heart of the Colorado River when he argued in favor of a high dam at Glen Canyon in 1954 and 1955. Wallace Stegner had told him early on that the beauty of Glen Canyon made Echo Park look almost ridiculous in comparison. In May 1954, when Governor J. Bracken Lee of Utah asked Brower whether he knew that a group of people was opposed to a dam in Glen Canyon for the same reasons that the Sierra Club was fighting the Echo Park dam, Brower had responded, "I know there are a group of people in Salt Lake who oppose Glen Canyon . . . [but] they are not the same reasons as mine. What we oppose is the invasion of a national park unit, and Glen Canyon is not in the National Park System." Brower recognized that since Glen Canyon was not specifically set aside and protected by law, a powerful argument for its preservation was missing from the equation. But Lee pressed his point. "Let me invite you to go through Glen Canyon," the Utah governor said to Brower, "and if you are a real lover of beauty, I think you will have to agree it is a far more attractive park than the Dinosaur National Monument; and I have seen them both." Brower, almost wistfully, replied, "If it were possible to save Glen Canyon and add it to the National Park System—but Glen Canyon is not part of the national park system and thus it is out of the purview of our stand. I don't think there will ever be any alternative found for the Glen Canyon reservoir. That is such an important part of the old Upper Colorado project, I don't see how even the nature-lovingest person of all could find a way to save that. I'm working on it."[14]

Brower eventually did cite some old Bureau of Reclamation reports on the geology of Glen Canyon to suggest that its Navajo sandstone was too porous to support a dam.[15] California Congressman Craig Hosmer, still hoping to protect his state's interests by defeating the entire Upper Colorado River Storage Project, picked up on Brower's argument. Hosmer traveled to Glen Canyon with a couple of independent geologists and collected samples of the Chinle shale which lies beneath the canyon's sandstone walls. On the day that the CRSP came up for vote in the House, Hosmer had one of his colleagues pour water into a glass containing a piece of the Chinle shale, which rapidly disintegrated. The message was graphic and ominous. But Arizona Congressman Stewart Udall, a staunch supporter of the Glen Canyon dam, had brought his own piece of stone, part of a core sample from the dam site. As he rose to respond to Hosmer, Udall dropped the small cylinder of sandstone into a glass of water. After describing the benefits of building the Glen Canyon Dam, Udall smiled and drank the clear water as his colleagues showered him with hearty guffaws. Before the day was over, the House approved the project.[16] David Brower sat in the House gallery and watched the vote go through.

Because the Echo Park dam had been deleted from the CRSP, most conservationists considered the House vote an unqualified victory. Brower himself got caught up in that mood for a while. When he saw Glen Canyon with his own eyes, however, and realized what was going to be flooded, Brower was suddenly filled with regret. "If I had *seen* it," he later said, "I would never have given up." Reflecting back on the day he had watched the House approve the CRSP and Glen Canyon Dam, Brower said, "The Sierra Club removed its opposition because it was in on the principle of protecting the national parks. My horrible mistake at that time was to have stayed in Washington, instead of to have grabbed the next plane back and called for an emergency meeting of the executive committee or the Sierra Club board to argue why we should have stayed in the battle. We should never have given away Glen Canyon or anything else until we knew more about what was there. I had at that point, I think, enough influence over the board that if I'd done that, Glen Canyon Dam would not have been built. I didn't make the right decision on that June [*sic*] day."[17]

Although many conservationists had second thoughts about sacrificing Glen Canyon, Brower took special umbrage at the dam. He sincerely believed—and still believes—that the dam's very existence is his fault. "It sounds quite fatuous, I'll grant you that," Brower admits, "but I was the one person who could have changed it. I keep flogging myself on that, and I guess I'll continue to."[18] With Glen Canyon weighing heavily on his conscience, Brower decided to focus his energies on a moving public eulogy to the place. Photographer Eliot Porter accompanied him on a raft trip through Glen Canyon, and later the two collaborated on an exhibit-format book called *The Place No One Knew: Glen Canyon on the Colorado*. Porter's striking photographs were accompanied by Brower's text. "Glen Canyon died in 1963 and I was partly responsible for its needless death," he confessed in the foreword. "Neither you nor I, nor anyone else, knew it well enough to insist that at all costs it should endure."[19] If the title was not quite accurate—after all, plenty of people knew about the beauty of Glen Canyon—it at least allowed Brower and other conservationists to deal better with the shame they felt at having given it away.

While they mourned the impending loss of Glen Canyon, conservationists had to swallow yet another bitter pill. The reservoir behind Glen Canyon Dam could potentially back water into Rainbow Bridge Natural Monument, a 160-acre preserve surrounding the world's largest natural stone arch. Although the authorizing legislation for Upper Colorado River Storage Project had explicitly dictated that "the secretary of the Interior shall take adequate protective measures to preclude impairment of Rainbow Bridge Natural Monument," Congress later refused to appropriate funds for a diversion dam to keep water out of the monument. If the water stored behind Glen Canyon Dam was allowed to encroach on Rainbow Bridge National Monument, conservationists would lose the principle of inviolability as well as Glen Canyon itself. What most conservationists had once considered an unqualified victory would seem hollow indeed.

Stewart Udall, the Arizona congressman who had been appointed secretary of Interior in 1960, was sensitive to the dilemma that confronted him at Rainbow Bridge National Monument. As a congressman, Udall had argued vehemently for Glen Canyon Dam, but he also appreciated the beauty of the slickrock canyon country and the value of undeveloped wilderness.[20] He could hardly order the Bureau of Reclamation to halt construction of a project on which it had already spent millions of dollars, nor could he allow the dam to stand unused once it had been completed. But he hoped to offset the damage to Rainbow Bridge National Monument by expanding it from a tiny 160-acre preserve to a real wilderness park. For good measure, he would also propose that national park status be granted to the area between the head of Glen Canyon Dam's reservoir—now called Lake Powell—and the junction of the Green and Colorado rivers. Conservationists would gain a great deal, Udall believed, if only they would compromise and allow a single slender fjord of Lake Powell to enter Rainbow Bridge National Monument.[21]

But by now, the word "compromise" caused many conservationists to cringe. David Brower, in particular, was disgusted with Udall's ambivalence, and he directed much of his anger at the amiable secretary of Interior. In the October 1961 issue of the Sierra Club *Bulletin,* Brower stated, "We now know that the life expectancy of one of America's greatest scenic resources, including the pristine approach to Rainbow Bridge, is reduced to fourteen months. The exact time is not important here. What needs to be chronicled is a flagrant betrayal, unequaled in the conservation history that sixty-eight years of Sierra Club *Bulletins* have recorded." Five months later, in an open letter to Udall carried by newspapers around the country, Brower observed caustically, *"Preclude* impairment, the law says. It doesn't say to plead excessive cost. Or to hustle through some kind of 'geological whitewash.' Or to arrange a series of show-me trips to lead editors and congressmen into believing that protection is just too much load on taxpayers and would tear up the countryside with roads and scars. . . . And when the law says preclude impairment, it spells it out in unmistakable words: 'no dam or reservoir . . . shall be within any national park or monument.' Not maybe. Not yes, but. Just *no.*"[22]

In the summer of 1962, on behalf of several allied conservation groups, the National Parks Association filed suit against Udall in U.S. District Court to forestall the closing of the diversion tunnels which allowed the Colorado River to bypass Glen Canyon Dam. The court dismissed the case on the ground that these groups did not have proper standing to sue. Early the next year, the Interior Department solicitor delivered to Udall his opinion that "the provisions originally included in the Colorado River Storage Project Act calling for protective measures at Rainbow Bridge National Monument have been suspended by the Congress and are no longer operative. Under the present state of the law applicable to Glen Canyon, it is the intent of Congress that construction and filling of the reservoir should proceed on schedule without awaiting the construction of barrier dams at Rainbow Bridge." With that opinion in hand, Udall decided to close the west diversion tunnel and begin to back water behind the Glen Canyon Dam.[23]

On the day which the diversion tunnel was scheduled to be closed, Brower flew to Washington in a desperate effort to persuade Udall not to shut the river off until adequate measures had been taken to protect Rainbow Bridge. Somehow Brower believed that an impassioned personal plea might convert Udall. But the secretary of Interior could not see Brower that morning because he and Floyd Dominy, commissioner of Reclamation, had arranged a press conference to announce plans for a whole new series of dams and diversion tunnels called the Pacific Southwest Water Plan. Instead of making his plea, Brower stood in the back of an auditorium at the Department of Interior and listened to the press conference, aghast. The multi-billion-dollar plan called for, among many other things, two new dams downstream from Glen Canyon. Brower could hardly believe what he was hearing—Dominy and the engineers at the Bureau of Reclamation wanted to dam the Grand Canyon.[24]

The Grand Canyon dams, which Dominy liked to call "cash registers," were similar to the Echo Park and Glen Canyon dams in that their primary purpose was to generate hydroelectricity. The Bureau knew the reservoirs would not conserve any water; in fact, in dry years, they would actually cause a net *loss* to the river through evaporation.[25] One dam was to be built in Marble Canyon, sixty miles downstream from the Glen Canyon Dam, and the other two hundred miles downstream at Bridge Canyon, near the point where the Colorado River flows into Lake Mead. Although neither dam would actually stand within the boundaries of Grand Canyon National Park, they would affect some forty miles of Grand Canyon National Monument and thirteen miles of the national park. Together they would flood 146 miles of the Colorado River and six hundred feet of much of the Grand Canyon.

As conservationists geared up for another protracted struggle against the Bureau of Reclamation, they could hardly help being cognizant of the lessons they had learned at Echo Park. Before Echo Park, many conservationists had been willing to compromise. In fact, in 1949 the Sierra Club Board of Directors had approved an earlier proposal to build the Bridge Canyon dam provided certain conditions were first met. David Brower was one of the board members who voted in favor of the compromise.[26] The loss of Glen Canyon changed Brower's attitude. "When they finally closed the diversion tunnels and the water began to rise, *then* it hurt. My God! Up until then I thought, well, maybe I can still do something about it."[27] Watching Glen Canyon fill with water tempered his resolve to protect the country's remaining wilderness. Never again would Brower allow politics, diplomacy, or compromise to interfere with that crusade.

Brower poured his zeal into the fight against the Grand Canyon dams. He followed up his book on Glen Canyon with another entitled *Time and the River Flowing,* with text written by François Leydet and photos by Philip Hyde and Martin Litton. In a sense, the book was a continuation of *The Place No One Knew.* "Each book," Brower wrote in the foreword:

> tells about the same extraordinary river and its great canyons. . . . Each book draws heavily upon perceptive interpretation by many of America's best writers of what these canyons mean to the world—what Glen Canyon could have meant and what Grand Canyon can always mean. Both books tell of the massive inflexibility and compulsive engineering that lost one canyon forever and seem determined to lose the other. Both books make the plea that this generation do better for all other generations than to let the Bureau of Reclamation carry out its present plans to destroy what is most important in Grand Canyon. The two books reinforce each other . . . to underline the tragedy it would be to let the Bureau of Reclamation repeat its mistake—not out of evil intent or incompetence, but from adamantly following a course of action that reveres engineering values and technology and ignores the human soul and sense of wonder.[28]

During the Echo Park controversy conservationists had learned that even the most eloquent emotional appeals could get lost in the favor-trading of special interest politics. Rather than rely solely on hortative pleas, conservationists marshaled experts to critique the economic and technical aspects of the proposed project.[29] They also rekindled the publicity campaign that they had used to such good effect in the Echo Park controversy. In March 1966, *Reader's Digest* ran an article denouncing the dams, and many other popular periodicals soon followed suit. "Right after the *Reader's Digest* article, *Life* ran a big goddamned diatribe," recalled Dan Dreyfus, a high-ranking official in the Bureau of Reclamation. "Then we got plastered by *My Weekly Reader.* You're in deep shit when you catch it from them. Mailbags were coming in by the hundreds stuffed with letters from schoolkids. I kept trying to tell Dominy that we were in trouble, but he didn't seem to give a damn."[30]

Indeed, although the publicity campaign was generating a lot of opposition to the dams, Dominy showed no signs of wavering. Brower decided that conservationists needed to pull out all the stops. On June 9, 1966, he spent some of the Sierra Club's discretionary fund to run two full-page ads that he devised with the help of San Francisco advertising executives Jerry Mander and Howard Gossage. The first, written by Brower, took the familiar form of an open letter addressed to Secretary Udall. The other, created by Mander, displayed the headline: "Now Only You Can Prevent the Grand Canyon from Being Flooded . . . For Profit." The ads were intentionally dramatic. "There's no point in writing an ad unless it's going to be talked about," Gossage explained to Brower. "It must be an event. It must *do* something."[31]

The ads did more than even Gossage anticipated. The day after they ran, the IRS delivered a message to the Sierra Club office in San Francisco informing its directors that the club's tax-deductible status was in jeopardy for lobbying. The warning cut off deductible contributions almost immediately, costing the club an estimated half million dollars. Brower believed the IRS action was orchestrated by a congressman who supported the dams. It was obviously a political strike, intended to sweep the Sierra Club from the public scene. It did exactly the opposite. Front-page articles in newspapers around the country and editorials in the *New York Times* and the *Wall Street Journal* denounced the IRS for its "assault on the right of private citizens to protest effectively against wrong-headed public policies" and for its "extraordinary departure from its snail's paced tradition." Membership in the Sierra Club surged as a result of the publicity. The media attention also inspired a brand new barrage of letters against the dams. As Brower later said, "People who didn't know whether or not they loved the Grand Canyon knew whether or not they liked the IRS."[32]

Although the Sierra Club Board of Directors was less than pleased that Brower had run the ad without bothering to consult them, they stood behind his decision. Encouraged by the board's tolerance, and unconvinced that the IRS action would hurt the club financially, Brower decided to disregard the pending risk of losing the club's tax-deductible status and ran another full-page ad. This one asked pointedly, "How can you guarantee these, Mr. Udall, if Grand Canyon is dammed for profit?" with the text of the ad listing development proposals that would encroach upon several additional units in the park system. A few weeks later, Brower ran yet another ad, this one asking caustically, "Should we also flood the Sistine Chapel so tourists can get nearer the ceiling?" The ads had the desired effect. By the end of 1966, the stream of angry letters arriving in the offices of key members

of the House and Senate Interior committees had turned into a tor-
rential flood of hundreds of *thousands* of letters a day. "I never saw
anything like it," recalled Dreyfus. "Letters were arriving in dump
trucks. Ninety-five percent of them said we'd better keep our mitts
off the Grand Canyon and a lot of them quoted the Sierra Club ads."[33]

The publicity campaign was clearly paying huge dividends, so
much so that in January 1967 the Bureau of Reclamation decided to
bring a compromise to the opening of the 90th Congress. They would
rename Bridge Canyon dam "Hualapai" as a token gesture to the
native people the dam was supposed to help and abolish Grand
Canyon National Monument to avoid infringement on a unit of the
national park system; in exchange, they would abandon the Marble
Canyon dam and expand Grand Canyon National Park. To the
Bureau's dismay, and the dismay of Arizona water boosters like Representative Morris Udall, conservationists refused to budge. Suppose
the Bureau built a "low, low, low Bridge Canyon dam, maybe one
hundred feet high," Udall asked Brower. "Is that too much? Is there
any point at which you compromise here?" Brower made clear that
there was not. "You are not giving us anything that God didn't put
here in the first place. . . . We have no choice. There have to be groups
who will hold out for those things that are not replaceable. If we
stop doing that, we might as well stop being an organization, and conservation organizations might as well throw in the towel."[34]

In February, while Dominy was out of the country, Secretary Stewart Udall withdrew his support of the Bridge Canyon dam and
instructed Assistant Secretary Ken Holum to replace it with an alternative that would stand a fair chance of passing Congress. Dominy
was furious, but there was little the Reclamation commissioner could
do about Udall's decision except sit and fume.[35] On August 8, 1967,
the Senate authorized the project without either dam, and a year later
the House did the same. Dominy later claimed that he abandoned the
dams not because he acknowledged the strong sentiment in favor of
preserving the Grand Canyon, but because "my secretary turned
chickenshit on me." Dreyfus's conclusion was a little different. "The
fact is we were licked. The conservationists and the press, and ultimately the public, licked the Bureau of Reclamation."[36]

Certainly, Udall's decision to abandon the Grand Canyon dams was
driven more by a realization that public attitudes and values had
changed than by a lack of personal courage. Although this new consciousness began to manifest itself in a variety of ways throughout the
1960s, it was perhaps nowhere more evident than in the legislation
adopted by Congress. The Wilderness Act of 1964 reserved over nine
million acres of undeveloped federal land. The Endangered Species
Act of 1966 established that preservation of critical habitat would take

priority over resource development. Both acts explicitly placed environmental objectives ahead of economic growth. The *coup de grace* for the Bureau of Reclamation came right on the heels of its defeat in the Grand Canyon. Only two days after the Bureau's watered-down Colorado River Basin Project Act became law, President Lyndon Johnson signed a bill establishing a national system of Wild and Scenic Rivers. "The established national policy of dam and other construction at appropriate sections of the rivers of the United States needs to be complemented," the law stated, "by a policy that would preserve other selected rivers in their free-flowing condition." The act instantly created eight components of the system, and by the late 1970s nineteen rivers—more than 1,600 miles—would receive similar protection.[37]

If it would be overly simplistic to say that Brower and other conservationists had single-handedly instigated this change in public attitudes, it would be just as wrong to say that they merely capitalized on a trend over which they had no control. The political successes of conservationists in the 1950s and 1960s perhaps relied upon a burgeoning "ecological consciousness" in the American public, but they also contributed to its growth. Brower and his contemporaries rode in the vanguard of a transformation of American values that gave rise to the modern environmental movement.

For the uncompromising attitude he brought to bear against the Bureau of Reclamation and other federal agencies, it is not unreasonable to think of David Brower as a "prophet" of the American conservation movement. He has been described by *Life* magazine, for instance, as America's "number one conservationist." In a series of articles for *The New Yorker* that later became an award-winning book, John McPhee referred to Brower as the "Archdruid" of American conservation.[38] Brower was, and to some extent still is, the movement's fulltime figurehead.

In biblical stories, the prophet who enlightens his people with a new message often receives a face full of stones as his thanks. Brower, too, has been pelted by stones. The uncompromising, confrontational tactics he used turned many of his colleagues against him—including a majority of the Sierra Club Board of Directors.

In May 1969, less than a year after the Grand Canyon battle came to a final close, the Sierra Club board gathered in San Francisco's Sir Francis Drake Hotel to decide whether to retain Brower as executive director. The board's discontent with Brower was fueled by many issues, but the most important was his seeming inability to be *reasonable*. Richard Leonard—a former president of the Sierra Club, the original sponsor for Brower's membership in the Sierra Club, and the nominator of Brower as executive director—expressed the board's view. "In the early years," Leonard stated, "Dave was absolutely

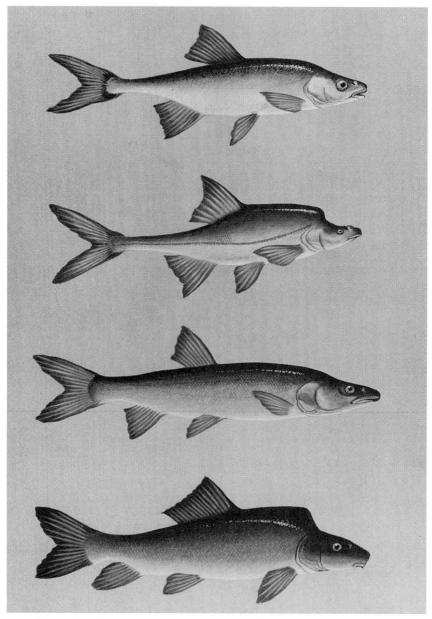

Over thousands of years, these native fish adapted to the swift, warm, muddy currents of the rivers of the Colorado River system. Dams blocked their spawning migration routes, cooled the water temperature, regulated river flows, and inundated needed habitat. The fish could not adapt to such radical changes and declined in number to their current endangered status. Efforts to save these fish, although heroic, may not be able to overcome the impacts of dams. *Dinosaur National Monument.*

magnificent as the leader of the club. He fought vigorously, aggressively, and—the point I want to emphasize—courteously. In later years, he started into his philosophy that Nice Nelly could never do the job. He impugned the motives of Forest Service people, Park Service people, congressmen. He seemed to feel that the end justified the means. The board passed resolutions insisting that he wage campaigns on demonstrable facts. Repeatedly, he has disregarded what the board has told him to do. He seems to think that it is he who knows what is best for the Sierra Club and for conservation in the long run, and that the board of directors is just standing in his way. The basis of his drive is that the earth is going to hell fast and something has to be done about it. Because of this, Dave will spend the resources of any organization he is with in unlimited fashion. 'We're not trying to save money, we're trying to save the world,' he will say, and then he will put thirty thousand dollars or so into another newspaper advertisement, without being authorized to do so by the board. . . . His ideals are good, but his naiveté would eventually destroy the organization. He believes that if he bankrupts the Sierra Club it is in a glorious cause."[39]

Brower still adamantly defends himself against charges of fiscal irresponsibility. He insists the money he spent on the Grand Canyon ads was justified, for instance, because "they [the board] authorized me to use the discretionary fund at my discretion." The loss of the club's tax status, he insists, was not an issue. "George Marshall I think was unhappy, and Richard Leonard was unhappy, but I don't know of any others who expressed unhappiness at the loss of the tax status." Brower believes that with very few exceptions, the board of directors agreed that "if we had to lose our tax status again to save the Grand Canyon, we'd lose it." Brower defended expenditures for the Sierra Club publishing program even more vehemently. "Our controller told me that if the present Sierra Club publishing efforts were accounted for the way *mine* were, instead of *losing* 250,000 dollars over that period, we'd have made a million. That was the kind of fiscal irresponsibility I was showing."[40]

It hardly matters now whether Brower was financially irresponsible or not; when gathered in San Francisco in 1969, the board of directors plainly believed that he was, and they were the final arbiters of the matter. When the meeting convened, everyone present knew it would end with Brower walking the plank. He opted instead to offer a last minute resignation. Martin Litton, Eliot Porter, Luna Leopold, and two other directors voted to keep Brower. Ten others voted to accept his resignation, including Richard Leonard and Ansel Adams.

By coincidence, a photograph Adams had taken many years before was featured on the front page of the *San Francisco Chronicle* that morning. The celebrated Wawona Tree, a giant sequoia in Yosemite National Park that had a roadway running through it, had toppled and shattered the previous winter. Adams had photographed the tree in better days, with a Pierce-Arrow nosing into the tunnel and a few people standing alongside. Besides Adams and Brower, few people in the room knew that one of those people was David Brower. The photo's caption read, "A Fallen Giant."[41]

The Consequences of Compromise

David Brower is now eighty-two years old. Since his bitter parting with the Sierra Club, he has been affiliated, in varying capacities, with other conservation organizations, including the John Muir Institute, Friends of the Earth, and the Earth Island Institute.[1] Although he never entirely eschewed politics, Brower has concentrated most of his efforts on educating people about environmental degradation and how to keep it from getting worse. He has a number of channels through which he can communicate that message, but he probably is most influential when speaking to people directly. For more than four decades, Brower has traveled around the United States and four continents explaining to whoever will listen that the world is going to hell fast and trying to inspire them to try to slow the rate of descent.

Brower is driven, at least in part, by the remorse he feels about Glen Canyon. His single-minded dedication to the conservation crusade reflects the depth of his contrition. He does not dwell somberly on Glen Canyon, yet he still speaks emotionally when he recalls the beauty of the place, or his own complicity in its destruction. Occasionally, when his smoldering grief is fanned by some vivid reminder of what was lost at Glen Canyon, his passion flares. Not long ago, for instance, Brower was in Montana to make a speech, which was to be followed by a film about Glen Canyon. Before rising to the podium, Brower became suddenly ill. He was ushered outside for some air, but refused a ride back to his hotel because he wanted to see the Glen Canyon film. Although he had seen it many times before, its effect was striking. "I watched that," Brower said, shaking his head, "and it totally recovered me. I was full of fire afterwards."[2]

Brower is in fact only one of many people who have taken personal umbrage at Glen Canyon Dam. While he may have special reason to feel guilty because of his pivotal role in the compromise which sacrificed it, almost everyone who experienced the beauty of Glen Canyon has shared his deep regret at what was lost. Wallace Stegner, writer and special advisor to Interior Secretary Stewart Udall, was among those mourners. Stegner had floated through Glen Canyon in 1947 while researching his definitive biography of John Wesley Powell. In 1954, when Brower recruited Stegner to edit and contribute to *This Is Dinosaur*, Stegner had warned him that it was a mistake to promote a high dam at Glen Canyon. He had seen much of the West, but in his opinion no place compared to that tranquil stretch of the Colorado. The scenery "is at once awesome and charming," Stegner wrote about Glen Canyon not long after his first visit. "The sheer cliffs of Navajo sandstone, stained in vertical stripes like a roman-striped ribbon and intricately cross-bedded and etched, lift straight out of the great river. . . . It is surely the handsomest of all the rock strata in this country." Yet Stegner knew from experience that the side canyons were even more enthralling. It was these narrow, sinuous chasms that hid unexpected pools and verdant pockets of fern and redbud, cottonwood and willow. Lake Powell covered most of these secret recesses with water as it rose, diminishing whatever scenery it did not inundate entirely. Stegner acknowledged that the lake would make the remaining side canyons "democratically accessible," but like most visitors who had been fortunate enough to see Glen Canyon, he considered the price unacceptably dear. "In gaining the lovely and the usable," he noted, "we have given up the incomparable."[3]

The West has had few residents as perceptive or as eloquent as Wallace Stegner. Undoubtedly, the memory of Glen Canyon colored his outlook on the region he called home and tempered his less-than-complimentary attitude toward the Bureau of Reclamation. Yet it was one of Stegner's students who would become the most scathing critic of Glen Canyon Dam. Edward Abbey was raised in Pennsylvania, but in spirit he was a native of the desert southwest even before he enrolled in the University of New Mexico in 1947, the year Stegner made his first trip through Glen Canyon. Abbey later won a Wallace Stegner Creative Writing Fellowship at Stanford, and before long he decided to see Glen Canyon for himself. In 1959, Abbey and a friend floated from Hite to Kane Creek landing, the last take-out point above the damsite, which was by then already swarming with construction workers. He wrote about the experience in a series of essays titled *Desert Solitaire*. "I was one of the lucky few who saw Glen Canyon before it was drowned," he told his readers. "In fact I only saw part of it but enough to realize that here was an Eden, a portion of the earth's original paradise." Abbey later worked as a seasonal park ranger on Lake Powell in the Glen Canyon National Recreation Area, where he experienced the area as it had been transformed by the dam. He left no doubt as to which he preferred. "The canyonlands did have a heart, a living heart, and that heart was Glen Canyon and the wild Colorado," Abbey concluded. "The difference between the present reservoir with its silent sterile shores and debris-choked side canyons, and the original Glen Canyon, is the difference between death and life. Glen Canyon was alive. Lake Powell is a graveyard."[4]

If David Brower was one of the first to lament what had been lost at Glen Canyon, Abbey was certainly one of the most strident. Their complaints were remarkably similar. Brower once expressed his contempt for the reservoir behind Glen Canyon Dam by stating, "The magic of Glen Canyon is dead. It has been vulgarized. Putting water in the Cathedral in the Desert was like urinating in the crypt of St. Peter's." Abbey agreed. "To grasp the nature of the crime that was committed," he wrote in *Desert Solitaire,* "imagine the Taj Mahal or Chartres Cathedral buried in mud until only the spires remain visible. With this difference: those man-made celebrations of human aspiration could conceivably be reconstructed while Glen Canyon was a living thing, irreplaceable, which can never be recovered through any human agency."[5] The fact that both Brower and Abbey equated Glen Canyon with a great cathedral is significant, because Glen Canyon had become for them a sacred space, a symbol of the value of wilderness and the destructiveness of industrial society grown out of control.

Not everyone held such contempt for Glen Canyon Dam, of course. Even before Lake Powell filled to capacity, the Glen Canyon National Recreation Area had become the most single most visited unit in the park system, outdrawing Yellowstone, Yosemite, even Grand Canyon. Of all the visitors who flocked to Lake Powell, no one admired it more than Floyd Dominy; the commissioner of Reclamation thought it the most beautiful lake anywhere on earth. Echoing Brower and Abbey, Dominy spoke of Glen Canyon—or, in this case, the reservoir that covered it—in vaguely religious terms. A Bureau publication entitled *Lake Powell: Jewel of the Colorado,* extolled the reservoir's transcendental virtues by stating, "To have a deep blue lake where no lake was before seems to bring man a little closer to God." Abbey scoffed at Dominy's self-gratulation. "In this case, Lake Powell, about 500 feet closer. Eh, Floyd?"[6]

At times it seemed Abbey's scorn knew no bounds. Brower may have pushed the limits of prudence while serving as the executive director of the Sierra Club, but he always maintained a certain degree of political acumen. As a writer with no board of directors peering over his shoulder and no political allies to alienate, Abbey was free from such constraints. The difference showed not only in his raw, unabashed style of writing, but in his presumed prescription for Glen Canyon. Whereas Brower had simply mourned Glen Canyon, Abbey dreamed—publicly—of redeeming it. In *Desert Solitaire,* Abbey imagined, almost wistfully, some "unknown hero with a rucksack full of dynamite" who would create the "loveliest explosion ever seen by man, reducing the great dam to a heap of rubble in the path of the river."[7] He picked up the same idea in 1975 in *The Monkey Wrench Gang,* a light-hearted book about a band of environmental saboteurs wreaking havoc in the desert southwest. When they weren't chopping down billboards, pulling up surveyors' stakes, and disabling bulldozers, Abbey's heroes dreamed of blowing up Glen Canyon Dam.[8] To many readers, Abbey's irreverent and confrontational prose seemed less a furtive complaint than a compelling call to action. His popularity grew steadily through the early 1970s, and by the end of the decade Abbey had joined Brower as one the figureheads of the American conservation movement; in the desert southwest, he was *the* figurehead.

Even as Abbey's popularity grew, however, the environmental movement itself was becoming gradually institutionalized and, in some ways, more moderate. The celebration of Earth Day in 1970 may have given environmentalism the appearance of a surging grassroots movement, but in the years that followed leaders of many conservation organizations were enticed into the government bureaucracy to staff new agencies such as the Environmental Protection Agency and to help implement clean air and water legislation. To the extent that these programs manifested the conservation movement's growing political presence, institutionalization was merely a sign that the movement had come of age. Yet maturity often brings disillusionment, and to many conservationists the political voice they had struggled to obtain did not seem loud enough to shout down the newer—and in many ways, more insidious—environmental threats of the 1970s. Problems such as overpopulation, air and water pollution, and deforestation could not be simply legislated out of existence. Although political action might help mitigate these problems in the short run, many conservationists believed it could only scratch the surface of a global environmental crisis driven by over-consumption and limitless economic growth. To all involved, it was clear that the environmental movement in the United States had fracturing along a new ideological fault line. The subtle fissure which separated conservationists following the Echo Park dam controversy had by the 1970s become a more formidable gulf, separating conservationists who focused on political reform from those who sought to reform the social values and ideals on which political institutions are based.[9]

The disillusionment which was spreading through the environmental movement seemed to reach critical proportions after the presidential election of 1980. The Reagan administration's general emphasis on deregulation and its particularly pointed attack on environmental regulations threatened to negate the incremental gains which mainstream conservationists had so painstakingly achieved over the previous two decades. After appointing blatant anti-environmentalists such as Anne Gorsuch and James Watt for key positions in his administration, Reagan orchestrated an executive realignment which virtually dismantled the Council on Environmental Quality, slashed the budget of the EPA, and hampered the implementation of environmental regulations through new administrative requirements.[10] Suddenly, conservationists seemed more vulnerable to the machinery of the political system than at any time in recent memory.

Ed Abbey had already suggested a more straightforward, and more satisfying, alternative to that political system. His call to action, resonating anew after Reagan's election, was not ignored. In 1981, several of Abbey's most ardent admirers decided to establish a self-proclaimed "radical" environmental organization called Earth First!. As a tribute to Abbey, the founders of Earth First! selected the monkey wrench as its trademark; as its motto and its philosophical underpinning, they chose the confrontational declaration, "No compromise in defense of Mother Earth." Abbey himself was on hand when, in the spring of 1981, more than seventy Earth First! activists gathered at Glen Canyon Dam for the group's first public demonstration. Avoiding the suspicious gaze of security personnel, a few activists sneaked away from the crowd and unfurled three hundred feet of tapering black plastic onto the face of the dam, creating the illusion of a giant crack. Ed Abbey's Monkey Wrench Gang had come to life.[11]

Over the last fifteen years, Earth First! has been one of the most visible elements in the radical environmental movement and is now well known for its direct action techniques of protest. Of course, it was no coincidence that Earth First! chose the ceremonial "cracking" of Glen Canyon Dam as its first public demonstration. The event was carefully planned to herald the arrival of a new breed of environmentalist, one who wants to remove conservation from the realm of political logrolling and compromise and place it in a higher, more sacred realm. To this type of conservationist, Glen Canyon Dam was a symbol of dual evils. It represented not only the industrial, growth-oriented society which was the source of their discontent, but also the dangers of pursuing reform of that society through conventional political action.

111

Today, almost forty years after the Echo Park was spared, there are still many people who wince at the mention of Glen Canyon Dam. Ed Abbey was not greatly exaggerating when he wrote, "In all of the Rocky Mountain, Inter-Mountain West, no man-made object has been hated so much, by so many, for so long, with such good reason, as that 700,000-ton plug of gray cement, blocking our river."[12] Many of the people who harbor these feelings are not inspired by fond memories of Glen Canyon, or haunted by images of its irrecoverable beauty—a good many hadn't been born when it was flooded. They despise the dam because it reminds them of the consequences of compromise. Engineers may have built the dam, but conservationists, beguiled by the political process, had given it their blessing.

It is important to realize that the radical element in the American conservation has real historical roots, that it is not a temporary fad. Indeed, the ideological chasm that has opened between radical and mainstream environmentalists reaches back to a physical chasm, a scenic canyon at the confluence of the Green and Yampa rivers called Echo Park.

Although the Echo Park dam controversy resolved internal tensions in the American conservation movement, it also created new ones. The fundamental dichotomy which plagued the movement since the days of Pinchot and Muir—whether conservation is ultimately concerned with managing resources for maximum efficiency or preserving nature for its aesthetic value—was resolved. After finding an effective political voice, conservationists successfully argued that preservation could in some cases be considered a kind of "wise use."

The Echo Park dam controversy also proved that conservationists had the political muscle to fight and win their battles in a political arena. For many conservationists the logical moral seemed to be that lobbying, negotiation, and compromise were the proper tools with which to fight future battles. Others, angered by the desecration of Glen Canyon and Rainbow Bridge, took from Echo Park exactly the opposite lesson. Inspired by charismatic leaders such as Brower and Abbey, these more radical environmentalists believed conservation problems were above redress except through sweeping changes in lifestyle and a repudiation of unlimited economic and industrial growth. The Echo Park controversy therefore represents a nascent battle among environmentalists as much as it does a struggle between conservationists and the Bureau of Reclamation. Despite the strong political coalition that conservationists forged to keep dams out of Dinosaur National Monument, the legacy of the Echo Park controversy is not one of lasting unity among conservationists; rather, it is a legacy of division.

Almost forty years after the fact, we are still struggling with the repercussions of the first great political deal cut by American conservationists.

The uncomfortable issue of compromise which grew out of the Echo Park dam controversy remains as troubling today as it was in the 1950s and 1960s. Whether the issue is logging in the Pacific Northwest, controlling industrial emissions of air and water pollutants, or the treatment and disposal of hazardous wastes, concerned individuals continue to debate the appropriateness of relying on negotiation and political trade-offs to resolve our environmental dilemmas. The compromises conservationists make on these issues will not be temporary paper agreements. They will in many cases create lasting, physical edifices— if not dams on rivers, then other irreversible additions, deletions, and modifications to the natural landscape.

Although there is yet no consensus on when compromise is acceptable and when it is not, even the most radical environmentalists acknowledge the need to wage battle in political trenches. David Brower, for instance, has noted that "politics is democracy's way of handling public business. There is no other. We won't get the kind of country in the kind of world we want unless people take a part in the public's business. Unless they embrace politics and people in politics."[13] George Sessions, one of the leading figures in the deep ecology movement, likewise acknowledged, "You've got to have legal machinery, you've got to have money and lobbying—all the work that the shallow environmentalists are doing. If we lost the shallow environmentalists overnight, we'd be in big trouble—these big corporations and agencies would roll over the environment in no time." Even Dave Foreman, one of the original founders of Earth First!, avers to be "the last one to say that electoral politics, court challenges, and lobbying for good legislation have no place in the tactics of our movement."[14]

But organizations which operate exclusively as political agents of change will remain unsatisfactory to those radical environmentalists who believe the answers to our environmental woes necessarily lie *outside* of the political system. Foreman, who has since left Earth First! to work on other projects, exemplifies this attitude. "Let's face it," he said recently, "our representative democracy has broken down. Our government primarily represents the big money boys and stacks the deck against reform movements. Playing only by the system's rules limits you. Trying to fit in, to not seem radical or extreme, to always seek compromise, keeps you pretty damn manageable."[15]

It is not entirely clear whether the environmental movement of the 1990s will be strengthened by its ideological diversity or impeded by it. There is the danger, of course, that the radicalism espoused by groups like Earth First! will alienate people who would otherwise be sympathetic to environmental goals, damaging the political credibility environmentalists have worked so hard to establish. Yet it is also plausible that a radical element will strengthen the environmental

An aerial view of Steamboad Rock showing the junction of the Yampa and Green rivers with the Mitten Park fault in the right background. *Photo by Jack Boucher, Dinosaur National Monument.*

movement by startling the public out its complacency, by discomfiting policymakers, and by forcing representatives of industry to deal in good faith with the mainstream conservation organizations. Perhaps the only certainty is that there will be loud and acrimonious debate among the conservationists who try to chart a common course on different ideological maps.

As the debate unfolds, conservationists must keep the lessons of the Echo Park dam controversy firmly in mind, for the conclusions they reach will in large measure determine in the quality of the landscape we leave to our descendants. By looking to the past, they might avoid mistakes in the future, mistakes which once made will not be easily undone, if they can be undone at all.

David Brower knows the pain and has gained the wisdom which come from bearing responsibility for such a mistake. In his old age, he can only hope for some unlikely redemption. Thinking of Glen Canyon, Brower recently said, "I want to see the dam collapse of its own weight. I don't want it blown up because that would be devastating all the way downstream. I'd like to see a valve inserted at river level and get it down to the dead storage level, and just stabilize it there. I want to see it recover. I've seen a lot of places [along the shore] where it had been a long time since [the water] had been up to high level, where you can see some recovery. Full recovery would take a few millennia, but I'd love to just see it start."[16]

In the meantime, Brower can take some comfort in knowing that it is still possible to walk along the high narrow ridge called Harper's Corner and look down several thousand feet at the Green and Yampa rivers flowing through Echo Park. From this distance Steamboat Rock seems almost fragile, as easily broken as a thin sliver of slate. The sheer-walled canyon that stretches around it is only part of a system of canyons winding their way to the horizon, but it is in the foreground and it is clearly the focal point of the viewscape. From this sweeping perspective Echo Park's beauty is austere, inaccessible, shrouded in silence. From the level of the river, the experience is quite different, and perhaps even more beautiful. Steamboat Rock, towering several hundred feet into the sky, seems anything but fragile. The place is alive, and everywhere there is evidence of its vital functions— tree limbs waving in the breeze, the darting flight of a canyon wren, an occasional bighorn clambering up a canyon wall, and always the sound of the living river.

Endnotes

Introduction

1 John Wesley Powell, *The Exploration of the Colorado River* (1875; abridged, Chicago: University of Chicago Press, 1957), 26.

2 Jack Sumner, *A Daily Journal of the Colorado Exploring Expedition*, copy in DNM Library; Powell, *The Exploration of the Colorado River,* 29–30.

3 *U.S. Statutes at Large* 39 (1916): 535. The language of the National Park Service Organic Act also has proved a constant challenge to the agency which must implement it.

4 Gifford Pinchot, *The Fight for Conservation* (Harcourt, Brace, 1910); [Robert Underwood Johnson], "The Neglect of Beauty in the Conservation Movement," *Century* (1910); John Muir, *Our National Parks* (1909).

5 Holway R. Jones, *John Muir and the Sierra Club* (San Francisco, Sierra Club Books, ca. 1965). For a complete account of the Hetch Hetchy controversy, see Stephen Fox, *John Muir and His Legacy: The American Conservation Movement* (Boston: Little Brown, 1981). Less complete treatment can be found in Nash, *Wilderness and the American Mind*, and Hays, *Conservation and the Gospel of Efficiency.*

6 Wallace Stegner, *The Sound of Mountain Water: The Changing American West* (Garden City, NY: Doubleday, 1969), 19.

7 Walter Prescott Webb, "The American West, Perpetual Mirage," *Harper's* (May 1957); Wallace Stegner, *The American West as Living Space* (Ann Arbor: The University of Michigan Press, 1987).

8 W. J. McGee, "Water as a Resource" (1909); John Widtsoe, "Success in Irrigation Projects" (1928). Both cited in Donald Worster, *Rivers of Empire: Water, Aridity, and the Growth of the American West* (New York: Pantheon, 1985), 127 and 188.

9 Bureau appropriation statistics cited in Worster, *Rivers of Empire,* 239; Henry Luce, "Endless Frontier," *Time* (July 30, 1951).

10 Marc Reisner, *Cadillac Desert: The American West and Its Disappearing Water* (New York: Viking Press, 1986), 125–27.

11 Reisner, *Cadillac Desert,* 126.

12 David Brower, testimony before House Committee on Interior and Insular Affairs, Subcommittee on Reclamation and Irrigation, January 26–27, 1954; E. O. Larson quoted in Howard Zahniser, "Shall We Dam Our National Park System?" Statement given to House Subcommittee on Irrigation and Reclamation, March 18, 1955, WSA box 17 mf 15.

13 Benton MacKaye to John Saylor, June 14, 1954, WSA box 33, mf 32.

14 John McPhee, *Encounters with the Archdruid* (New York: Farrar, Straus, & Giroux, 1971), 164.

Chapter One: Growth of a Dinosaur

1 Dinosaur's 209,000 acres dwarf Bryce Canyon (35,835 acres) and Arches (73,234 acres), and exceed Zion's 147,035 acres.

2 The Taylor Grazing Act of 1934, which provided for federal retention and regulation of the unreserved portions of the public rangelands, was the single most important piece of legislation in Roosevelt's program for federal stewardship of the public domain. See E. Louise Peffer, *The Closing of the Public Domain: Disposal and Reservation Policies 1900–1950* (Stanford: Stanford University Press, 1951).

3 In "Address of the Honorable Harold L. Ickes," transcribed remarks in "Proceedings of the National Parks Superintendents Conference," November 19, 1934. Cited in T.H. Watkins, *Righteous Pilgrim: The Life and Times of Harold L. Ickes 1874–1952* (New York: Henry Holt and Company, 1990), 550.

4 Between 1933 and 1937, Ickes and Roosevelt set aside or enlarged several national monuments in the desert Southwest, including Cedar Breaks, Death Valley, Saguaro, White Sands (1933), Joshua Tree (1936), and Capitol Reef (1937).

5 Eliot Blackwelder, "Geological Exhibit" in Wallace Stegner (ed.), *This Is Dinosaur: Echo Park Country and Its Magic Rivers* (New York: Alfred Knopf, 1955), 18–30.

6 Wallace Stegner, "The Marks of Human Passage" in *This Is Dinosaur,* 3–17; Frank B. Sarles, "History of Dinosaur National Monument" (dissertation, Dinosaur National Monument Library, 1969), 1–9.

7 The two most thoroughly documented expeditions are those of General William Ashley, who navigated a portion of the upper reaches of the Green in 1825, and William Manly, who did the same in 1849.

8 John Wesley Powell, *Report on the Arid Regions of the United States,* ed. Wallace Stegner (1878; reprint, Cambridge: Harvard University Press, 1962).

9 Wallace Stegner, introduction to Powell, *The Exploration of the Colorado River* (1875; abridged, Chicago: University of Chicago Press), xii and xvii. For a more complete biography of Powell's life and work, see Stegner, *Beyond the Hundredth Meridian: John Wesley Powell and the Second Opening of the West* (Boston: Houghton Mifflin, 1954).

10 Herbert E. Evison, NPS "Report of Inspection Trip to Proposed Yampa Canyon Area," Hoops file, DNM, July 5–13, 1935.

11   Elmo Richardson, *Dams, Parks, and Politics: Resource Development and Preservation in the Truman–Eisenhower Era* (Lexington: University of Kentucky Press, 1973), 48. Excerpts from the Federal Power Commission's decision are quoted in a confidential memorandum from Reclamation Commissioner Michael Straus to Secretary of Interior Oscar Chapman, February 24, 1950, DNM files.

12   Don Hatch (Vernal, Utah) to author, December 11, 1992; Richardson, *Dams, Parks, and Politics,* 49.

13   *Salt Lake Tribune*, September 4, 1938; Richardson, *Dams, Parks, and Politics,* 49. Madsen described his action in a signed affidavit which was widely distributed by proponents of the Echo Park dam, DNM files. Ickes's instructions to Park Service personnel are cited in the "Memorandum on Echo Park Dam Controversy in Dinosaur National Monument," August 18, 1954, WSA box 12, mf 11.

14   Executive Order, quoted in "Shall We Dam Our National Park System?" Howard Zahniser to House Subcommittee on Irrigation and Reclamation, March 28, 1955, WSA box 17, mf 15.

15   See Federal Water Power Act (41 Stat. 758) and amendments (41 Stat. 1353 and 49 Stat. 838).

16   Howard W. Baker, resident landscape architect, "Report of Inspection Trip to Proposed Yampa Canyon Area," Hoops file, DNM, June 9–19, 1935.

17   C. Wirth to T. Moskey, ca. October 1937. Cited in Richardson, *Dams, Parks, and Politics,* 49.

Chapter Two: "Here Comes a Boom"

1   Memorandum from project superintendent to regional director, cited in "Bureau of Reclamation Activities in Dinosaur National Monument," DNM files.

2   The Reclamation Service did conduct one of its first projects in north-central Utah. Surveys for Strawberry Reservoir began in 1903, and the project was completed in 1913. For a brief history of irrigation in Utah through the 1950s, see George Clyde, "History of Irrigation in Utah," *Utah Historical Quarterly,* vol. 27 (1959).

3   There are several books which tell the story of the Colorado River Compact. The most complete is Norris Hundley Jr.'s *Water and the West: The Colorado River Compact and the Politics of Water in the American West* (Berkeley: University of California Press, 1975). Less complete treatment can be found in Philip Fradkin, *A River No More,* Marc Reisner, *Cadillac Desert,* and Donald Worster, *Rivers of Empire.*

4   NPS memorandum from Assistant Superintendent McLaughlin, January 6, 1939; Bureau of Reclamation memorandum from chief engineer Walter, January 9, 1939; NPS memorandum to the director, January 19, 1939. All cited in "Bureau of Reclamation Activities in Dinosaur National Monument," DNM files.

5   Mather cited in Michael Frome, *Regreening the National Parks* (Tucson: University of Arizona Press, 1992), 46–7. See also Tom Turner, *Sierra Club: 100 Years of Protecting Nature* (New York: Abrams in association with the Sierra Club, 1991), 120.

6 Harold L. Ickes to J. Horace McFarland, November 17, 1937, cited in Shivers-Culpin, *Putting the Secretary on the Hot Seat* (August 1991); Ickes, statement made November 12, 1937, cited in Watkins, *Righteous Pilgrim* (New York: Henry Holt, 1990), 582.

7 A. Demaray to J. McLaughlin, February 18, 1939; D. Canfield to Region II, August 1, 1939. Cited in "Bureau of Reclamation Activities in Dinosaur National Monument," DNM files.

8 D. Canfield to A. Demaray, February 14, 1940; Memorandum from Acting Chief Engineer Harper, March 29, 1940. Cited in "Bureau of Reclamation Activities in Dinosaur National Monument," DNM files.

9 When Drury accepted the appointment of Park Service director, Robert Sterling Yard, a founding member of the Wilderness Society, called him "one of the highest-minded conservationists in the United States." Cited in Watkins, *Righteous Pilgrim,* 579.

10 Quoted by Dave Brower, interview with author, August 11, 1994.

11 Worster, *Rivers of Empire,* 239–40. See also Donald C. Swain, "The Bureau of Reclamation and the New Deal," *Pacific Northwest Quarterly,* vol. 61 (1970).

12 Frome, *Regreening the National Parks,* 61.

13 Cited in Richardson, *Dams, Parks, and Politics,* 50.

14 Memorandum of understanding, November 4, 1941. Cited in "Bureau of Reclamation Activities in Dinosaur National Monument," DNM files.

15 Commissioner of Reclamation Page to National Park Service Director Drury, March 3, 1942; director to commissioner, March 12, 1942, DNM files.

16 *Federal Register,* July 13, 1943, 10370–71.

17 The oversight is mentioned in a letter from Ira Gabrielson to Fred Packard, September 14, 1950, WSA box 33, mf 17.

18 National Park Service Director Drury to Commissioner of Reclamation Page, December 1, 1943. Cited in "Bureau of Reclamation Activities in Dinosaur National Monument," DNM files.

19 Cited in Richardson, *Dams, Parks, and Politics,* 51.

20 For more detail on the growing concern for wilderness between the world wars, see Nash, *Wilderness and the American Mind,* 182–209. Also useful is John Henneberger, *Chronology of Events Relating to Wilderness Preservation to 1960,* WSA box 29, mf 21.

21 Leopold, "Wilderness as a Land Laboratory," *Living Wilderness* 6 (1941): 3. Cited in Nash, *Wilderness and the American Mind,* 198.

22 "Conservationists See Nothing 'Mysterious' in Ousting of Drury," *Deseret News,* February 17, 1951.

23 *Vernal Express,* July 7, 1948.

24 *Utah: A Guide to the State,* compiled by workers of the Writers' Program of the Works Progress Administration for the State of Utah (New York: Hastings House, 1941), 7.

25 Wallace Stegner, *The Gathering of Zion,* 2.

26 Worster, *Rivers of Empire,* 76–77; George Clyde, "History of irrigation in Utah," *Utah Historical Quarterly,* vol. 27 (1959).

27 *Utah: A Guide to the State,* 8; "Central Utah Project," pamphlet distributed by Utah State Water and Power Board, WSA.

28 *Utah: A Guide to the State,* 3–4; Wallace Stegner, *Mormon Country* (New York: Duell, Sloan, and Pierce, 1942), 28.

29 Don Hatch to author, December 11, 1992.

30 Reported in *Vernal Express,* March 21, 1946. The *Vernal Express* ran its first story on the Echo Park and Split Mountain dams more than four years earlier (December 11, 1941) but did not report regularly on the project until the war was over.

31 *Vernal Express,* March 28, 1946, and July 7, 1948.

32 "Central Utah Project," pamphlet distributed by Utah State Water and Power Board,WSA.

33 See, for example, pamphlet entitled "Utah's Last Waterhole," WSA.

34 Richardson, *Dams, Parks, and Politics,* 22.

35 *Vernal Express,* April 1 and 11, 1946.

36 Bureau of Reclamation, *A Comprehensive Report on the Development of Water Resources,* 80th Congress, 1947, House Document 419.

37 *Vernal Express,* December 4, 1947, and April 6, 1949.

38 Harry S. Truman, statement, Phoenix, Arizona, September 24, 1948, *Public Papers,* 566–67; Harry S. Truman to Irving Brant, February 20,1951, cited in Richardson, *Dams, Parks, and Politics,* 66.

39 G.F. Ingalls, "Summary of Findings Regarding Possible Road Locations and Connections," July 30, 1943, DNM files.

40 Confidential memorandum for the director from regional director II, October 7, 1943, DNM files.

41 Draft of the "NPS Survey of the Recreational Resources of the Colorado River Basin," August 1944, DNM files; see also citations in memorandum from commissioner of Reclamation to secretary of Interior, December 20, 1949, DNM files.

42 National Park Service Director Drury to Secretary of Interior Ickes, September 6, 1945, DNM files.

43 Ickes later said of the memorandum of understanding, "The worst of this surprising surrender on Mr. Drury's part was that this agreement should have been . . . brought to my attention, since I was Secretary at the time. . . . If I had continued as Secretary of the Interior, I would have requested Mr. Drury's resignation long ago in no uncertain terms." H. Ickes to editor, *New York Times,* February 15, 1951, DNM files.

44 *Vernal Express,* July 7, 1948.

45 Memorandum from assistant secretary of Interior to commissioner of Reclamation, June 14, 1949, cited in "Bureau of Reclamation Activities in Dinosaur National Monument," DNM files.

46 NPS memorandum from Acting Director A. E. Demaray to regional directors I and II, July 1949; Memorandum from acting director, NPS, to commissioner of Reclamation, August 16, 1949, DNM files.

47 Roy Webb, *Riverman: The Story of Bus Hatch* (Rock Springs, WY: Labyrinth Publishing Co., 1989), 86.

48 NPS memorandum from superintendent, Rocky Mountain National Park, to regional director II, September 29, 1949, DNM files.

49 *Vernal Express,* September 21 and 28, 1949; Krug cited in *Living Wilderness* (Winter 1948–49).

50 NPS memorandum from superintendent of Dinosaur National Monument to superintendent of Rocky Mountain National Park, ca. September 1949, DNM files.

51 NPS memorandum from superintendent of Dinosaur National Monument to superintendent of Rocky Mountain National Park, December 6, 1949; NPS memorandum from regional director II to regional director III, August 29, 1949, DNM files.

52 "Resume of Region II Files Relating to Proposed Split Mountain and Echo Park Units, Colorado River Storage Project," DNM files.

53 NPS memorandum from National Park Service director to regional director II, October 4, 1949; Memorandum National Park Service director to secretary of Interior, November 18, 1949; Memorandum from David Canfield, superintendent of Rocky Mountain National Park, to Parley R. Neeley, area engineer of Bureau of Reclamation, December 9, 1949, DNM files.

54 Ickes cited in memorandum from Newton Drury to secretary of the Interior, September 6, 1945, DNM files.

55 Drury cited in Richardson, *Dams, Parks, and Politics,* 63, and Dave Brower, *Environmental Activist, Publicist, and Prophet,* (Bancroft Library Oral History Program at University of California, Berkeley), 52.

Chapter Three: A House Divided

1 See Stegner, Wallace, *Beyond the Hundredth Meridian: John Wesley Powell and the Second Opening of the West* (Boston; Houghton and Mifflin, 1954). Stegner later claimed that Powell would have been opposed to the Echo Park dam.

2 Mather cited in Frome, *Regreening the National Parks,* 211.

3 Ickes, "Farewell Secretary Krug," *New Republic,* November 28, 1949.

4 Cited in Richardson, *Dams, Parks, and Politics,* 56.

5 Chapman, address to American Planning and Civic Association, January 19, 1950, cited in *Living Wilderness* (Winter 1949–50): 28–29.

6 For Ickes's desire to create "Department of Conservation," see T. H. Watkins, *Righteous Pilgrim,* 484–94; Chapman to A. Carhart, August 3, 1950, Carhart Papers, box 74; Calhoun cited in Shivers-Culpin, *Putting the Secretary on the Hot Seat,* 1.

7 Confidential memorandum from commissioner of Reclamation to secretary of Interior, February 24, 1950, DNM file.

8 Memoranda from National Park Service director to secretary of the Interior, November 18 and December 30, 1949; Confidential memorandum from National Park Service director to secretary of the Interior, March 3, 1950, DNM files.

9 Reisner, *Cadillac Desert,* 139.

10 Stringham cited in *Vernal Express,* March 2, 1950.

11 *Vernal Express,* January 26 and February 2, 1950; Richardson, *Dams, Parks, and Politics,* 57.

12 Chapman, introductory statement in "Shall Dams Be Built in Dinosaur National Monument?" (excerpts from the testimony to the secretary of Interior, April 3, 1950), partial transcript available in Carhart Papers.

13 The arguments in favor of Echo Park dam are made succinctly in the statements of N. B. Bennett, Jr., assistant director, Branch of Project Planning, Bureau of Reclamation, and Clifford H. Stone, director, Colorado Water Conservation Board, in "Shall Dams Be Built in Dinosaur National Monument?" (excerpts from the testimony to the secretary of the Interior, April 3, 1950). The bureau's position is also developed in a report by Michael W. Straus, commissioner, and E. O. Larson, director, Region Four, Bureau of Reclamation, "Brief Report on the Importance of the Echo Park and Split Mountain Units, Colorado River Storage Project and Their Relation to the Existing Dinosaur National Monument," December 1949, copy, DNM files.

14 Ibid.

15 Shivers-Culpin, *Putting the Secretary on the Hot Seat,* 19; "Oral History Interview with Oscar Chapman," transcript, 461–62.

16 Gabrielson, opening statement in "Shall Dams Be Built in Dinosaur National Monument?" (excerpts from the testimony presented to the Hon. Oscar L. Chapman, secretary of Interior, at public hearing held April 3, 1950), 10. All subsequent statements cited in this chapter are from same source.

17 "Shall Dams Be Built in Dinosaur National Monument?" George W. Kelley, 27–28; Joseph Penfold, 57; Howard Zahniser, 73A; U.S. Grant, 50; Kenneth D. Morrison, 69–70.

18 "Shall Dams Be Built in Dinosaur National Monument?" Olmstead, 19; Gabrielson, 11; Ickes, 81A.

19 "Shall Dams Be Built in Dinosaur National Monument?" Ickes, 81B; Grant, 49; Voigt, 62; Harold E. Anthony, 79.

20 "Shall Dams Be Built in Dinosaur National Monument?" Albright, 48; Gabrielson, 11; B. Robinson, 32; Penfold, 59.

21 "Shall Dams Be Built in Dinosaur National Monument?" Grant, 49–56; Zahniser, 73.

22 "Shall Dams Be Built in Dinosaur National Monument?" (introduction to transcript by Newton Drury, April 27, 1950).

Chapter Four: The Strength of Convictions

1 *Vernal Express,* April 13, 1950.

2 *Vernal Express,* June 1, 1950.

3 Drury, introduction to "Shall Dams Be Built in Dinosaur National Monument?" (excerpts from the testimony presented to to the Hon. Oscar L. Chapman, secretary of the Interior, April 27, 1950).

4 Glenn Sandiford, *Echo Park: The Forgotten Contribution of Bernard DeVoto* (unpublished paper written for SUNY College of Environmental Science and Forestry, 1988), 7.

5 *Salt Lake Tribune,* June 28, 1950.

6 Joint memorandum from secretary of Interior to commissioner, Bureau of Reclamation, and the National Park Service director, July 27, 1950, DNM files.

7 Oscar Chapman to Arthur Carhart, August 3, 1950, Carhart Papers, box 74.

8 *Vernal Express,* August 31, 1950.

9 Arthur Carhart to Oscar Chapman, September 10, 1950, Carhart Papers, box 74.

10 B. DeVoto to H. Frank, July 29, 1950; H. Ickes to B. DeVoto, December 12, 1950. Cited in Richardson, *Dams, Parks, and Politics,* 58.

11 Cited in Shivers-Culpin, *Putting the Secretary on the Hot Seat,* 19–20.

12 William Voigt to Joe Penfold and Arthur Carhart, December 11, 1950, Carhart Papers.

13 DeVoto, "Shall We Let Them Ruin Our National Parks?" *Saturday Evening Post,* July 22, 1950.

14 For examples of other articles, see Arthur Carhart, "The Menaced Dinosaur National Monument," *National Parks Magazine* (January–March 1952); Devereux Butcher, "This Is Dinosaur," *National Parks Magazine* (October–December 1950), and "In Defense of Dinosaur," *Audubon Magazine* (July 1952); U.S. Grant, III, "Dinosaur Dams Are Not Needed," *Living Wilderness* (Autumn 1950); Margaret Murie, "A Matter of Choice," *Living Wilderness* (Autumn 1950); Philip Hyde, "Nature's Climax at Dinosaur," *Living Wilderness* (Autumn 1952).

15 H. Albright to B. DeVoto, July 21, 1950, cited in Sandiford, *The Forgotten Contribution of Bernard DeVoto,* 6.

16 "Conservationists See Nothing 'Mysterious' in Ousting of Drury," *Deseret News,* February 17, 1951, copy DNM files; *Denver Post,* July 22, 1950; Straus, cited in "Let's Be Fair, Mike," *Nature Magazine* (October 1951).

17 See DeVoto, "Easy Chair: The West Against Itself," *Harper's* (January 1947).

18 *Denver Post,* August 9, 1950.

19 Watkins quoted in *Vernal Express,* January 4, 1951; the authorization bills introduced to 82d Cong., 1st sess., were H. R. 8980 (Granger), H. R. 9053 (Bosone), and S. 3839 (Thomas).

20 Richardson, *Dams, Parks, and Politics,* 62

21 Chapman later admitted that he offered Drury the position of special assistant simply to remove him from the daily work of the Interior Department. "He wouldn't have anything to do with administering," Chapman reflected, "except what I handed to him each day, and that would be nothing." "Oral History Interview with Oscar Chapman," 460–61.

22 Editorial, *New York Times,* February 14, 1951; *Washington Post,* February 17, 1951.

23 Leland quoted in *Living Wilderness* (Spring 1951), 41; Sandiford, *The Forgotten Contribution of Bernard DeVoto,* 8; and Richardson, *Dams, Parks, and Politics,* 67 and 69.

24 "Conservationists See Nothing 'Mysterious' in Ousting of Drury," *Deseret News,* February 17, 1951, copy, DNM files; *Vernal Express,* February 15, 1951.

25 H. Ickes to editor, *New York Times,* February 15, 1951; H. Ickes to Charles Sauers, February 13, 1951, DNM files.

26 "Oral History Interview with Oscar Chapman," transcript, 460–62; DeVoto to Eugene Lee, May 17, 1951, cited in Sandiford, *The Forgotten Contribution of Bernard DeVoto,* n. 50, 27.

27 Headline in *Vernal Express,* January 18, 1951.

28   *Vernal Express,* February 15, 1951.

29   Reported in *Living Wilderness* (Spring 1951): 34.

30   Open letter from John G. Will, secretary and general counsel, Upper Colorado River Commission, March 5, 1954, Carhart Papers, box 350; Oscar Chapman to B. Frank Ward, November 7, 1952, DNM files.

31   See, for example, the official comments and recommendations of the state of Colorado, returned to Chapman in June 1951, which emphasized "the need for immediate authorization and construction" of the Echo Park dam, Carhart Papers, box 67A.

32   Cited in Shivers-Culpin, *Putting the Secretary on the Hot Seat,* 21–22.

33   Memorandum from Conrad Wirth to Oscar Chapman, cited in Shivers-Culpin, *Putting the Secretary on the Hot Seat,* 22.

34   Chapman's remarks to National Audubon Society, quoted in *Vernal Express,* November 22, 1951.

35   Wirth's visit to Vernal reported in *Vernal Express,* July 17, 1952; Telegram from Vernal Chamber of Commerce to Secretary of Interior Oscar Chapman, July 23, 1952, DNM files; Open letter from B. Frank Ward to Oscar Chapman, printed in *Vernal Express,* September 10, 1953.

36   Chapman to A. Watkins, November 21, 1951, cited in *Living Wilderness* (Winter 1951–52): 31. Chapman's letter to Congress cited in *Living Wilderness* (Winter 1952–53): 23.

37   Straus to Chapman quoted in *Vernal Express,* February 8, 1951; B. Frank Ward to Oscar Chapman, September 21, 1951, Hoops File, DNM.

38   Truman's remarks reported in *Denver Post,* October 8, 1952.

39   Eisenhower campaign statements cited in Richardson, *Dams, Parks, and Politics,* 74.

40   Watkins quoted in *Vernal Express,* February 1, 1951.

41   DeVoto, "Easy Chair: The West Against Itself," *Harper's* (January 1947).

Chapter Five: Voices for the Wilderness

1   Richardson, *Dams, Parks, and Politics,* 82.

2   Jean Begeman, "McKay: Where Ickes Stood," *New Republic,* December 8, 1952.

3   McKay speech, Montesano, Washington, June 23, 1951, cited in Richardson, *Dams, Parks, and Politics,* 84

4   Morse cited in Richardson, *Dams, Parks, and Politics,* 86

5   J. B. Wood, "Seedbeds of Socialism," *Nation's Business* (October 1950).

6   D. Eisenhower, press conference statement, June 17, 1953; T. Dewey to D. McKay, December 20, 1952. Cited in Richardson, *Dams, Parks, and Politics,* 119 and 85.

7   Ralph Tudor, "I'm Glad I Came to Washington," *Saturday Evening Post,* November 27, 1954.

8   Cited in Richardson, *Dams, Parks, and Politics,* 97.

9   Straus cited in Reisner, *Cadillac Desert,* 154 and 144–45.

10   M. Straus to D. McKay, February 9, 1953, cited in Richardson, *Dams, Parks and Politics,* 95.

11   Richardson, *Dams, Parks, and Politics,* 96.

12   For a complete history of the Sierra Club, see Tom Turner, *Sierra Club: 100 Years of Protecting Nature* (New York: Abrams, 1991), or Michael Cohen, *History of the Sierra Club 1892–1970* (San Francisco: Sierra Club, 1988).

13   Leonard's and Hyde's trips to Dinosaur are mentioned in *The Living Wilderness* (Autumn 1950 and Autumn 1952, respectively). Letter from Richard Leonard to congressmen, May 7, 1952, DNM files; Frome, *Regreening the National Parks,* 214.

14   Brower, *Environmental Activist,* transcript, 111.

15   Turner, *Sierra Club,* 131.

16   Webb, *Riverman,* 93.

17   From the question period following Brower's testimony to the House Subcommittee on Irrigation and Reclamation, January 26–27, 1954 (excerpts of this testimony can be found in Dave Brower, *For Earth's Sake: The Life and Times of David Brower* [Salt Lake City: Peregrine Smith, 1992], 334–40). Also see Steve Bradley, "Folboats Through Dinosaur," *Sierra Club Bulletin* (December 1952).

18   David Brower, interview with author, August 11, 1994.

19   John Muir, *Our National Parks* (1901; San Francisco: Sierra Club, 1991); Muir, "The Tuolumne Yosemite in Danger," *Outlook,* November 2, 1907.

20   Charles Eggert to author, March 22, 1993. Also see Brower, *Environmental Activist,* transcript, 111–13.

21   Martin Litton, "Children in Boats Run Utah Rapids," Los Angeles *Times,* August 30, 1953; "Shooting Rapids in Dinosaur," *National Geographic* (March 1954).

22   Richard Leonard to Bus Hatch, July 31, 1953, Don Hatch personal file.

23   *Vernal Express,* June 18, 1953; Brower to William B. Wallis, June 24, 1953, Hoops file, DNM.

24   Brower to L. Y. Siddoway, June 25, 1953, Hoops file, DNM.

25   Don Hatch to author, December 11, 1992.

26   *Vernal Express,* January, 11, 1951. See also Webb, *Riverman,* 92.

27   Bus Hatch to Jess Lombard, March 9, 1954, Don Hatch personal file; Don Hatch to author, December 11, 1992.

28   J. Penfold to C. Edward Graves, September 12, 1955, WSA box 33, mf 17.

29   D. Brower to J. Lombard, October 21, 1953, DNM files, L2415.

30   J. Penfold to C. Edward Graves, September 12, 1955, WSA box 33, mf 17.

31   "Tomorrow's Playground for Millions of Americans," Upper Colorado River Commission, WSA box 33, mf 32.

32   McKay speech to Los Angeles Chamber of Commerce, February 3, 1954, DNM files. For more on Mission 66, see Frome, *Regreening the National Parks,* 64–66.

33   Fred Packard to Alfred Knopf, February 14, 1955, cited in Richardson, *Dams, Parks, and Politics,* 113.

34   Zahniser, Howard, "Shall We Dam Our National Park System?" statement delivered to House Subcommittee on Irrigation and Reclamation, March 28, 1955, WSA box 17, mf 15.

35 Brower, statement to Associated Sportsmen of California, September 17, 1954, DNM files; Brower, testimony before House Subcommittee on Irrigation and Reclamation, January 26–27, 1954.

36 Quoted by Brower in testimony to House Subcommittee on Irrigation and Reclamation, January 26–27, 1954.

37 Arthur Carhart, press release, ca. February 1954, Carhart Papers, box 74.

38 B. Robinson to J. Marr, July 20, 1953; B. Robinson to R. Leonard, July 23, 1953, Hoops file, DNM.

39 D. Brower to "Cooperators," September 8, 1953; R. Leonard to J. Marr, August 31, 1953, Hoops file, DNM.

40 B. Robinson to R. Leonard, July 23, 1953, Hoops file, DNM.

41 Richardson, *Dams, Parks, and Politics,* 102

42 B. DeVoto to Hugh Butler, December 1, 1952; DeVoto to H. Chollis, April 10, 1953, cited in Richardson, *Dams, Parks, and Politics,* 101, 105.

43 McKay speech to Los Angeles Chamber of Commerce, February 3, 1954, DNM files.

44 Ralph Tudor, *Notes,* May 10, 1953, cited in Richardson, *Dams, Parks, and Politics,* 154.

45 *Vernal Express,* March 28, 1946.

46 McKay to Wesley D'Ewart, June 29, 1953, cited in Richardson, *Dams, Parks, and Politics,* 104.

47 Richardson, *Dams, Parks, and Politics,* 98.

48 Open letter from John G. Will, secretary of the Upper Colorado River Commission, March 5, 1954, Carhart Papers, box 350; D. McKay to D. Eisenhower, December 10, 1953, cited in Richardson, *Dams, Parks, and Politics,* 133.

49 A. Carhart to B. DeVoto, March 11, 1954, Carhart Papers, box 83.

50 McKay quoted in *Vernal Express,* February 18, 1954.

51 *The Living Wilderness* (Autumn 1953): 29; Packard, "A Call to Battle," *National Parks Magazine* (January–April, 1953).

52 Carhart, unfinished drafts of letters, December 1953, Carhart Papers, box 83.

Chapter Six: "No Holds Are Barred"

1 Pughe, quoted in "Notes on Meeting of Colorado Water Conservation Board," transcribed by M.E. Murie, October 14, 1954, Carhart Papers, box 350; Arthur Carhart, "Last Chance to Save Echo Park," ca. June 1955, WSA box 33, mf 27.

2 Brower, *Environmental Activist,* transcript, 116.

3 Ralph Tudor, testimony before House Subcommittee on Irrigation and Reclamation, January 18, 1954.

4 Brower, testimony to House Subcommittee on Irrigation and Reclamation, January 26–27, 1954. A transcript of the testimony with notes written by Brower soon afterwards can be found in DNM files. Excerpts of the testimony and question period which followed can also be found in Brower, *For the Earth's Sake,* 326–340.

5 Leopold to Brower, copy, no date, DNM files. Leopold apparently spoke with Brower shortly after and made the same point in plainer language.

"He told me, 'Stick to your birdwatching,'" Brower recalled with a laugh (Brower, interview with author, August 11, 1994).

6   Brower, interview with author, August 11, 1994.

7   Brower, testimony to House Subcommittee on Irrigation and Reclamation, January 26–27, 1954. Also see Brower, *Environmental Activist,* transcript, 118 and 132–33, and *For the Earth's Sake,* 334.

8   Looking back on that afternoon, Brower himself discredited that theory. "No, I think they adjourned because it was getting late," he later said, laughing. "That was my recollection." *Environmental Activist,* transcript, 117.

9   Brower, testimony before House Subcommittee on Irrigation and Reclamation, January 26–27, 1954.

10   Ibid.

11   Open letter from John G. Will, secretary of the Upper Colorado River Commission, March 5, 1954, Carhart Papers, box 350; *Vernal Express,* January 28, 1954.

12   Tudor to Subcommittee Chairman William Harrison, March 9, 1954, and May 14, 1954; Howard Zahniser to Ralph Tudor, May 20, 1954; "Interior Department Urged to Abandon Echo Park Dam," press release, the Wilderness Society, WSA box 34, mf 2.

13   Zahniser to Tudor, May 20, 1954, WSA box 34, mf 2.

14   Tudor to Harrison, May 14, 1954, WSA box 34, mf 2.

15   Brower, *Environmental Activist,* transcript, 117; Turner, *Sierra Club,* 145.

16   *Vernal Express,* February 18, 1954; Joseph Dodge, director, Bureau of the Budget to D. McKay, March 18, 1954, Carhart Papers, box 74; White House Press Release, May 20, 1954, Carhart Papers, box 83; *Vernal Express,* March 21, 1954 .

17   *Vernal Express,* February 18, 1954. The optimism in Utah had been building for months. When fifteen members of the House Interior Committee had visited Dinosaur in September 1953, the *Vernal Express* declared in a bold headline, "Echo Park Dam Site Acclaimed by Congressional Committee" (September 17, 1953).

18   D. Brower to Sherman Adams, May 25, 1954, copy in DNM files.

19   Fred Packard to D. Brower, June 17, 1954, WSA box 33, mf 17.

20   Ibid.

21   Brower, interview with author, August 11, 1994.

22   *U.S. v. Harriss;* Tom Turner, *Sierra Club: 100 Years of Protecting Nature,* 148.

23   Sierra Club, *Bulletin* (February1954); Brower, *Environmental Activist,* 128.

24   Wallace Stegner (ed.), *This is Dinosaur: Echo Park Country and Its Magic Rivers* (New York: Alfred Knopf, 1955). Also see Brower, *Environmental Activist,* transcript, 128.

25   Brower, address at 30th annual convention of Associated Sportsmen of California, Stockton, California, September 17, 1954, DNM files.

26   Nash puts the ratio of letters opposed to the dam at eighty to one, *Wilderness and the American Mind,* 216. Letter to Adams cited in Richardson, *Dams, Parks, and Politics,* 144.

27 Interestingly, some conservationists who saw this film believed it was a more potent statement *against* the dam and suggested trying to obtain a copy. NPS memorandum from David Canfield to director, region II, Hoops file, DNM; *Vernal Express,* March 6, 1952, and March 18, 1954.

28 Pamphlet by Upper Colorado River Commission, *Conquering Our Frontiers Through Development of the Upper Colorado River Basin,* WSA; Pamphlet, *Straight Talk from the Aqualantes,* copy in Hoops file, DNM; Statement of Leslie A. Miller to House Committee on Interior and Insular Affairs, March 18, 1955, Carhart Papers, box 350.

29 *Vernal Express,* November 4 and December 23, 1954.

30 *Straight Talk from the Aqualantes,* copy in Hoops file, DNM.

31 Joe Penfold, "Report on Governor's Meeting" (Cheyenne, Wyoming, January 4, 1955), Carhart Papers, box 67a; The office of Representative John P. Saylor, "By Any Other Name Echo Park Is the Same," press release, January 20, 1955, WSA box 33, mf 30.

32 One vivid exception to this rule is a pamphlet entitled "Conquering Our Frontiers," printed and distributed by the Upper Colorado River Commission, WSA. The imagery of the pamphlet—showing railroads, covered wagons, pioneer families—supports its message: "The conquering of modern frontiers by developing the latent land and water resources of . . . the Upper Colorado River Basin can be expected to yield benefits of national significance."

33 Statement of Herbert F. Smart to House Subcommittee on Irrigation and Reclamation, March 16, 1955, cited in a press release from the office of Representative Craig Hosmer, "An Engine Without Pistons," February 15, 1956, WSA box 33, mf 30; pamphlet, "Wildlife Experts on Echo Park Dam," WSA box 33, mf 32.

34 Watkins speech in Salt Lake City, June 5, 1954, cited in memorandum from H. Zahniser to D. Brower, June 14, 1954.

35 Richardson, *Dams, Parks, and Politics,* 135–36.

Chapter Seven: Strange Bedfellows

1 HR. 270 (sponsored by Dawson of Utah), HR. 2836 (sponsored by Fernandez of New Mexico), and HR. 3383 and HR. 3384 (both sponsored by Wayne Aspinall of Colorado).

2 *Congressional Record,* 84th Cong., 1st sess., 101, pt. 10: 12220–21.

3 Pamphlet, "Straight Talk from the Aqualantes"; Bernard DeVoto to Arthur Carhart, March 8, 1954, Carhart Papers.

4 Brower, *Environmental Activist,* transcript, 122.

5 Joe Penfold, "Report on Governor's Meeting" (Cheyenne, Wyoming, January 4, 1955), Carhart Papers, box 67a; "Pres. Eisenhower and Gov. Ed Johnson on Big Dam Foolishness," April 7, 1955, copy in WSA box 33, mf 17; *Salt Lake Tribune,* January 9, 1955; *Vernal Express,* January 6, 1955.

6 Leslie A. Miller, testimony to House Subcommittee on Irrigation and Reclamation, March 18, 1955, Carhart Papers, box 350; Angrilantes, "Fool Dams and Purple Potatoes," newsletter, April 18, 1955, Hoops file, DNM.

7 John Saylor to D. McKay, March 6, 1954, cited in Richardson, *Dams, Parks, and Politics,* 140.

8 Moley's objections to the CRSP are most vividly presented in "Perspective: Irrigation—Hydropower's Expensive Partner," *Newsweek* (May 17, 1954). Also see Moley, "Perspective: What Price Reclamation?" *Newsweek* (February 14, 1955).

9 Elmer Peterson, *Big Dam Foolishness: The Problem of Modern Flood Control and Water Storage* (New York: Devin-Adair, 1954); "Pres. Eisenhower and Governor Ed Johnson Offer Views on 'Big Dam Foolishness,'" reprinted from "Conservation on the March," *Pine Cone* (April 7, 1955), WSA box 33, mf 17.

10 DeVoto to John F. Kennedy, September 15, 1954, cited in Richardson, *Dams, Parks, and Politics,* 139; *New York Times,* April 24, 1955.

11 Leslie Miller, "The Battle That Squanders Billions," *Saturday Evening Post,* May 14, 1949; statement to House Committee on Interior and Insular Affairs, March 18, 1955, Carhart Papers, box 350; "Dollars into Dust," *Reader's Digest* (May 1955).

12 Angrilantes newsletters, "Fool Dams and Purple Potatoes," April 18, 1955, and "Frying-Pan Arkansas Boondoggle Project," April 23, 1955, DNM files.

13 Leslie A. Miller, statement before House Subcommittee on Irrigation and Reclamation, March 18, 1955, Carhart Papers, box 350. Brower, in a memorandum entitled "Summary of Statement in Support of Dinosaur National Monument," March 17, 1955, put the figure $1460 per acre. Other sources put it as high as $1,750 (Reisner, *Cadillac Desert,* 148).

14 The LaPlata project in Colorado would cost $947 per acre, and the Navajo Project in New Mexico an estimated $1540 per acre. Figures are those given by Leslie A. Miller in his testimony to the House Subcommittee on Irrigation and Reclamation, March 18, 1955.

15 Miller, testimony to the House Subcommittee on Irrigation and Reclamation, March 18, 1955; Miller cited in memorandum prepared by National Wildlife Federation and state delegates, March 12, 1955, WSA box 17, mf 9.

16 The actual cost of the Missouri Valley Project had increased 3.6 times over the original estimate. Brower, "Summary of Statement in Support of Dinosaur National Monument," March 17, 1955, Carhart Papers.

17 William Voigt, "Recommendations of the Izaak Walton League of America to the President's Water Resources Policy Commission," March 22, 1950; director, Bureau of the Budget, to secretary of Interior, March 18, 1954, Carhart Papers, box 74; Miller, testimony to House Subcommittee on Irrigation and Reclamation, March 18,1955.

18 Lee's comments, made during the question period following Brower's testimony to Hoover Commission Task Force on Water and Power in May 3, 1954, DNM files.

19 Zahniser to McKay, June 8, 1954, WSA box 33, mf 17; McKay to Zahniser, June 15, 1954, WSA Box 34, mf 6.

20 Fred Packard, "Memorandum to Members of Congress," February 11, 1954, WSA box 33, mf 17.

21 Bestor Robinson, statement in "Shall Dams Be Built in Dinosaur National Monument?" (excerpts from the testimony to the secretary of the Interior, April 3, 1950).

22 Exchange between Brower and Lee made during the question period

following Brower's testimony to Hoover Commission Task Force on Water and Power, May 3, 1954, DNM files.

23 Pamphlet, "Sound Development and Unimpaired Parks: A Way to Have Both," Sierra Club, ca. June, 1955; Brower, *Environmental Activist,* transcript, 118.

24 John Baker to Howard Zahniser, April 12, 1955, WSA box 33, mf 27; in response to "Shall We Dam Our National Park System?" Zahniser's statement to House Subcommittee on Irrigation and Reclamation, March 28, 1955, WSA box 17, mf 15.

25 Fred Smith to various conservationists, May 27, 1955, WSA box 34, mf 2; Olaus Murie, "Plea for the Green and Yampa River Canyons," testimony to House Subcommittee on Irrigation and Reclamation, January, 1954, WSA box 33, mf 30.

26 *Congressional Record,* 84th Cong., 1st sess., 101, no. 4, 4806 and 4813; Also see "Dinosaur Journal," newsletter published by the Council of Conservationists, ca. June 1955, DNM files.

27 "Dinosaur Journal," newsletter published by the Council of Conservationists, ca. June 1955, DNM files; Fred Smith to various conservationists, May 20, 1955, WSA box 34, mf 2.

28 "Dinosaur Journal," newsletter published by the Council of Conservationists, ca. June 1955, DNM files.

29 Open letter from the Council of Conservationists, DNM files; Arthur Carhart, "Last Chance to Save Echo Park," ca. June 1955, copy in WSA box 33, mf 27.

30 Dawson cited in Nash, *Wilderness and the American Mind,* 218, and in an open letter from the Council of Conservationists, DNM files.

31 *Denver Post,* October 31, 1955; Resolution cited in letter from Senator Clinton P. Anderson to Fred Smith, November 3, 1955, WSA box 33, mf 32.

32 Craig Hosmer, press release, "An Engine Without Pistons," February 15, 1956, WSA box 33, mf 30.

33 Senator Clinton P. Anderson to Fred Smith, November 3, 1955, WSA box 33, mf 32.

34 43 USC 620b. Relevant excerpts from the Upper Colorado River Storage Act can be found in confidential memorandum entitled "The Dinosaur Park Dilemma: An Analysis," prepared by D. Brower, September 23, 1957, Carhart Papers.

Chapter Eight: The Place No One Knew
1 Russell Martin, *A Story That Stands Like a Dam: Glen Canyon and the Struggle for the Soul of the West* (New York: Henry Holt and Company, 1989), 86–87.

2 Quote attributed to Governor Simpson of Wyoming, cited in letter from Ira Gabrielson to Arthur Watkins, May 7, 1956, WSA box 34, mf 3.

3 *Salt Lake Tribune,* March 2, 1956.

4 Lombard to Brower, March 15, 1956, DNM files.

5 H.R. 10636 (Saylor) and H.R. 10614 (Aspinall).

6 Watkins quoted in *Salt Lake Tribune,* April 19, 1956; Gabrielson to

Watkins, May 7, 1956; Telegram from D. Brower to A. Watkins, cited in letter from Watkins to Gabrielson, April 24, 1956, WSA box 34, mf 3.

7 Frank E. Moss to Don Hatch, November 3, 1959, WSA box 33, mf 27; Clyde quoted in *Salt Lake Tribune*, August 23, 1958.

8 Relevant excerpts from Allott's bill (S. 2577) are found in "Dinosaur Proposed Park," part of the stenographic transcript from hearings held before the Senate Committee on Interior and Insular Affairs, July 8, 1958, WSA box 33, mf 18; Also valuable is a confidential memorandum entitled "The Dinosaur Park Dilemma: An Analysis," prepared by David Brower, September 23, 1957, Carhart Papers.

9 D. Brower to H. Albright, August 27, 1954; H. Albright to H. Zahniser, July 21, 1957; Memorandum from Fred Smith "To All Conservationists," August 14, 1957, WSA box 33, mf 27.

10 Brower, confidential memorandum entitled, "The Dinosaur Park Dilemma: An Analysis," September 23, 1957, Carhart Papers; Fred Packard to Secretary of Interior Fred Seaton, August 15, 1957; Memorandum from Olaus J. Murie to members of the National Parks Association Board of Trustees, August 12, 1957, WSA box 34, mf 3.

11 J. Saylor to H. Albright, August 10, 1959, WSA box 33, mf 27.

12 Olaus Murie to the National Parks Association Board of Trustees, August 12, 1957, WSA box 34, mf 3.

13 Brower, *Environmental Activist,* transcript, 90, 154, and 209.

14 Brower, *Environmental Activist,* transcript, 140; for a statement on the beauty of Glen Canyon predating the passage of the Upper Colorado Storage Project Act, see "Rainbow Land of Glen Canyon," *Natural History* (June 1950); Exchange between Lee and Brower during question period of Brower's testimony before the Water Resources and Power Task Force of the Commission on Organization of the Executive Branch of the Government, May 3, 1954, DNM files.

15 Brower testimony before Senate Subcommittee on Irrigation an Reclamation, February 28 thru March 4, 1955.

16 The demonstration on the floor of the House is described in Martin, *A Story That Stands Like a Dam,* 70–72.

17 Brower, interview with author, August 11, 1994; Brower, *Environmental Activist,* transcript, 130 and 140.

18 Brower, interview with author, August 11, 1994.

19 Eliot Porter, *The Place No One Knew: Glen Canyon on the Colorado,* edited by David Brower (San Francisco: Sierra Club, 1963).

20 For Udall's views on wilderness generally, see Udall, "To Save the Wonder of the Wilderness," *New York Times Magazine,* May 27, 1962.

21 Martin, *A Story That Stands Like a Dam,* 219–220.

22 *Sierra Club Bulletin,* October 1961 and March 1962.

23 Martin, *A Story That Stands Like a Dam,* 238–39.

24 Ibid.

25 Reisner, *Cadillac Desert,* 283.

26 Brower, *Environmental Activist,* transcript, 112.

27 Brower, interview with author, August 11, 1994.

28 François Leydet, *Time and the River Flowing: Grand Canyon,* edited by David Brower (San Francisco, Sierra Club Books, 1964).

29 Brower, *Environmental Activist,* 149–50; Reisner, *Cadillac Desert,* 295–96.

30 Dreyfus cited in Reisner, *Cadillac Desert,* 299.

31 Brower, *Environmental Activist,* 146–47.

32 Martin, *A Story That Stands Like a Dam,* 271–73; Brower, *Environmental Activist,* transcript, 146–48 and 150–53. Copies of the Grand Canyon advertisements can be found in the appendix of Brower, *Environmental Activist.* Brower notes that the directors "had gone through this all six years earlier" when they established the Sierra Club Foundation, "to be ready if we should lose our tax-deductible status." Deductibility, Brower points out, "was important for a very small fraction of the club's income—big grants and estates. No one cared whether the then $9 dues were deductible. Subsequent ads pointed out that because of the IRS action contributions would not be deductible. They came in regardless, and the membership grew rapidly." Letter to author, January 15, 1995.

33 Dreyfus cited in Reisner, *Cadillac Desert,* 297.

34 Exchange between Udall and Brower cited in Nash, *Wilderness and the American Mind,* 232–33, and discussed in Brower, *Environmental Activist,* transcript, 151.

35 Dan Dreyfus was the person who informed Dominy of the project amendment. "I thought he was going to go through the roof," Dreyfus said of the encounter, "but Dominy always had a way of catching you off guard. His reaction was complete and total lack of interest. He already knew all about it. He just said, 'I don't even want to hear about it,' and told me to get the hell out of his office. He didn't even look up from what he was reading on the desk." (Reisner, *Cadillac Desert,* 290).

36 Dominy and Dreyfus cited in Reisner, *Cadillac Desert,* 262 and 300.

37 Nash, *Wilderness and the American Mind,* 236. Also see "Preserving Our Wild and Scenic Rivers," *National Geographic,* July 1977.

38 McPhee, *Encounters with the Archdruid.*

39 Cited in McPhee, *Encounters with the Archdruid,* 215–16.

40 Brower, interview with author, August 11, 1994.

41 McPhee, *Encounters with the Archdruid,* 217–18.

Epilogue: The Consequences of Compromise

1 When enough time had passed to heal old wounds, Brower even renewed his affiliation with the Sierra Club. Brower received the John Muir award, the club's highest honor, and he currently serves as an honorary vice president.

2 Brower, interview with author, August 1994.

3 Wallace Stegner, *The Sound of Mountain Water,* 117, 123–24, 128.

4 Edward Abbey and Philip Hyde, *Slickrock,* 64 and 66.

5 Brower cited in McPhee, *Encounters with the Archdruid,* 240; Edward Abbey, *Desert Solitaire* (New York: Ballantine, 1968), 174.

6 Walter Edwards, "Lake Powell: Waterway to Desert Wonders," *National Geographic,* July 1967, 44–75; Dominy is widely acknowledged as

the primary author of "Lake Powell: Jewel of the Colorado" (Washington DC: Department of Interior); Abbey and Hyde, *Slickrock,* 67

7 Abbey, *Desert Solitaire,* 174 and 188.

8 Abbey, *The Monkey Wrench Gang* (Philadelphia, Lippencott, 1975).

9 The distinction between shallow and deep ecologists was made as early as 1973; see Arne Naess, "The Shallow and the Deep, Long-Range Ecology Movement: A Summary."

10 Gorsuch was appointed director of the Environmental Protection Agency and Watt served as secretary of Interior.

11 The Earth First! demonstration at Glen Canyon Dam is described in many different sources, including Martin, *A Story That Stands Like a Dam;* Manes, *Green Rage: Radical Environmentalism and the Unmaking of Civilization;* and Susan Zakin, *Coyotes and Town Dogs: Earth First! and the Environmental Movement.*

12 Edward Abbey, *Down the River* (New York: Penguin 1982), 186.

13 Dave Brower, Introduction to Hugh Nash, editor, *Progress as if Survival Mattered* (San Francisco: Friends of the Earth, 1977).

14 Sessions quoted in Kirkpatrick Sale, "The Forests for the Trees: Can Today's Environmentalists Tell the Difference?" *Mother Jones,* November, 1986. Foreman quoted in Steve Chase, editor, *Defending the Earth: A Dialogue between Murray Bookchin and Dave Foreman* (Boston: South End Press, 1991).

15 Foreman quoted in Steve Chase, editor, *Defending the Earth: A Dialogue between Murray Bookchin and Dave Foreman* (Boston: South End Press, 1991), 66–67.

16 Brower, interview with author, August 1994. Brower has pitched the idea to Dan Beard, the current Reclamation commissioner. "He told me in his office that he wants no more dams, and would even like to take some down," Brower says.

Notes on Sources

Historical Collections

The archives at Dinosaur National Monument contained a wide variety of materials related to the Echo Park controversy, including correspondence between National Park Service officials, conservationists, and other officials in the Department of Interior. Mr. Herm Hoops, Green River District Naturalist for the National Park Service, also made available his own personal file, including a number of relevant primary documents which were salvaged from the flooded basement of a park service employee. The archives at Dinosaur National Monument are abbreviated as "DNM files" in the endnotes. Mr. Hoops's personal file is referred to as "Hoops file, DNM."

The historical archives of The Wilderness Society, kept under the supervision of the Denver Public Library's Department of Western History, were also invaluable. In addition to providing a glimpse into the Wilderness Society, the archives contained correspondence generated by other organizations such as the Sierra Club and the National Parks Association, and a large collection of pamphlets and other propaganda used in the Echo Park controversy. The Wilderness Society's archives are abbreviated as "WSA" in the footnotes, followed by box and folder number.

Oral History and Personal Interviews

David R. Brower—Environmental Activist, Publicist and Prophet, a series of interviews conducted with David R. Brower, by Susan Schrepfer, 1974–1978. Bancroft Library Oral History Program at University of California, Berkeley.

Oral History Interview with Oscar Chapman, Harry S. Truman Library, Independence, Missouri, June 1980. [Excerpts on file in historical archives at Dinosaur National Monument].

Personal interviews and written correspondence with: Don Hatch (Vernal, Utah) August and December 1992, April 1993; Charles Eggert (Rhinebeck, New York) March 1993 and August 1994; David Brower (Boston, Massachusetts) August 11–12, 1994.

Personal Papers

The Arthur Carhart Papers, kept at the Denver Public Library's Department of Western History, contained valuable correspondence that provided a regional perspective on the conservationists' campaign to save Dinosaur. The public papers of Presidents Truman and Eisenhower, which have been published in book form, also proved valuable.

Government Documents

Bureau of Reclamation. *The Colorado River: A Comprehensive Report on the Development of Water Resources.* House Document 419, 80th Congress. Washington: Department of Interior, 1947.

Bureau of Reclamation. *Project Feasibilities and Authorizations.* Washington: Department of Interior, 1949.

Bureau of Reclamation. *Project Feasibilities and Authorizations.* Washington: Department of Interior, 1957.

National Park Service. *A Survey of the Recreational Resources of the Colorado River Basin.* Washington: Department of Interior, 1950.

U.S. Department of Interior. *Proceedings Before the United States Department of the Interior: Hearing on Dinosaur National Monument, Echo Park and Split Mountain Dams,* April 3, 1950. [A partial transcript of these hearings is available in the Arthur Carhart Papers. A complete transcript can be found at the Department of Interior Library, Washington, D.C.]

U.S. Congress, Senate, Committee on Interior and Insular Affairs, Subcommittee on Irrigation and Reclamation. *Hearings: Colorado River Storage Project,* 83d Cong., 2d Sess. June 28–July 3, 1954.

U.S. Congress, House, Committee on Interior and Insular Affairs, Subcommittee on Irrigation and Reclamation. *Hearings: Colorado River Storage Project,* 83d Cong., 2d Sess. January 18–23, 25, 28, 1954.

U.S. Congress, Senate, Committee on Interior and Insular Affairs, Subcommittee on Irrigation and Reclamation. *Hearings: Colorado River Storage Project,* 84th Cong., 1st Sess. February 28, March 1–5, 1955.

U.S. Congress, House, Committee on Interior and Insular Affairs, Subcommittee on Irrigation and Reclamation. *Hearings: Colorado River Storage Project,* 84th Cong., 1st Sess. Part 1: March 9, 10, April 18, 20, 22, 1955; Part 2: March 11, 14, 16–19, 28, 1955.

Periodical Literature

A trip through the *Reader's Guide to Periodical Literature* for the years 1949–1957 turned up over two hundred articles related to the Dinosaur controversy in popular periodicals. Some articles were written by leading players in the Dinosaur battle, and these had special significance. For instance, Secretary of Interior Doug McKay often defended his department directly through the media, and the highly regarded conservationist Bernard DeVoto exercised considerable influence through his column the *Easy Chair*. Membership bulletins put out by conservation groups were especially valuable. I relied most heavily on the Wilderness Society's *Living Wilderness* (predecessor of *Wilderness* magazine), but also referred less frequently to the *Sierra Club Bulletin, Audubon* magazine, and *National Parks Magazine*.

Newspaper accounts were also an important source. Because I referred to so many articles, they cannot be listed separately here, although the more important articles are cited in the footnotes. I generally relied on the *New York Times* to provide an eastern perspective and the *Denver Post* for a western perspective. The Wilderness Society and Dinosaur National Monument archives both contained a number of clippings from papers such as *Salt Lake Tribune, Washington Post, Los Angeles Times,* and *San Francisco Chronicle*. Clippings from the *Vernal Express,* on file in the archives at Dinosaur National Monument, provided a more regional perspective.

Historical Journals, Periodical Literature, and Unpublished Papers

Clyde, George. "History of irrigation in Utah." *Utah Historical Quarterly,* Vol. 27 (1959).

Henneberger, John W. "Chronology of Events Relating to Wilderness Preservation to 1960." (Unpublished paper) August, 1960. [Copy available in the archives of The Wilderness Society, Western History Department, Denver Public Library].

McCloskey, Michael. "Wilderness Mmovement at the Crossroads." *Pacific Historical Review,* Vol. 41 (1972).

Parfit, Michael. "Earth First!ers Wield a Mean Monkey Wrench." *Smithsonian Magazine,* Vol. 21 (April, 1990).

Sale, Kirkpatrick. "The Forest for the Trees: Can Today's Environmentalists Tell the Difference?" *Mother Jones,* Vol. 11 (November 1986).

Sandiford, Glenn. "Echo Park: The Forgotten Contribution of Bernard DeVoto." (Unpublished Paper written for SUNY College of Environmental Science and Forestry) 1988. [Copy available in Hoops file at Dinosaur National Monument].

Shivers-Culpin, Mary. *Putting the Secretary on the Hot Seat: The Bureau of Reclamation and the National Park Service at Odds.* (Address given to meeting of Western History Association in Austin, Texas) August, 1991.

Stegner, Wallace. "It All Began with Conservation." *Smithsonian Magazine,* Vol. 21 (April, 1990).

Swain, Donald. "The Bureau of Reclamation and the New Deal." *Pacific Northwest Quarterly,* Vol. 61 (1970).

Toney, Sharon. *Conservationists' Role in the Echo Park Dispute.* (Unpublished paper written for Environmental History Program at University of Texas) April, 1990.

Books

Abbey, Edward. *Desert Solitaire: A Season in the Wilderness*. New York: Ballantine Books, 1968.

Abbey, Edward. *The Monkeywrench Gang*. Philadelphia: Lippencott, 1975.

Abbey, Edward. *Down the River*. New York: Penguin, 1982.

Abbey, Edward, and Philip Hyde. *Slickrock*. San Francisco: Sierra Club Books, 1971.

Brower, David. *For Earth's Sake: The Life and Times of David Brower*. Salt Lake City: Peregrine Smith Books, 1992.

Chase, Steve (ed.). *Defending the Earth: A Dialogue between Murray Bookchin and Dave Foreman*. Boston: South End Press, 1991.

Cohen, Michael P. *History of the Sierra Club 1892–1970*. San Francisco: Sierra Club, 1988.

DeVoto, Bernard. *The Easy Chair*. Boston: Houghton Mifflin, 1955.

Dunlap, Riley E. and Angela D. Mertig (eds). *American Environmentalism: The U.S. Environmental Movement, 1970–1990*. Philadelphia: Taylor and Francis, 1992.

Fradkin, Phillip. *A River No More: The Colorado River and the West*. New York: Alfred A. Knopf, 1981.

Frome, Michael. *Regreening the National Parks*. Tucson: University of Arizona Press, 1992.

Hays, Samuel P. *Conservation and the Gospel of Efficiency: The Progressive Conservation Movement 1890–1920*. Cambridge: Harvard University Press, 1959.

Hays, Samuel P. *Beauty, Health, and Permanence: Environmental Politics in the United States 1955–1985*. New York: Cambridge University Press, 1987.

Hundley, Norris, Jr. *Water and the West: The Colorado River Compact and the Politics of Water in the American West*. Berkeley: University of California Press, 1975.

Ise, John. *Our National Park Policy: A Critical History*. Baltimore: Johns Hopkins University Press, 1961.

Leydet, François. *Time and the River Flowing: Grand Canyon*. Edited by David Brower. San Francisco: Sierra Club Books, 1964.

Manes, Christopher. *Green Rage: Radical Environmentalism and the Unmaking of Civilization*. Boston: Little, Brown and Company, 1990.

Martin, Russell. *A Story That Stands Like a Dam: Glen Canyon and the Struggle for the Soul of the West*. New York: Henry Holt and Company, 1989.

McPhee, John. *Encounters with the Archdruid*. New York: Farrar, Straus and Giroux, 1971.

Muir, John. *Our National Parks*. Reprinted from the First Edition of 1909. San Francisco: Sierra Club Books, 1991.

Nash, Roderick F. *Wilderness and the American Mind*. New Haven: Yale University Press, 1967.

Nash, Roderick F. (ed.). *American Environmentalism: Readings in Conservation History*. Third Edition. New York: McGraw-Hill, 1990.

Peffer, E. Louise. *The Closing of the Public Domain: Disposal and Reservation Policies, 1900–1950*. Stanford: Stanford University Press, 1951.

Peterson, Elmer T. *Big Dam Foolishness: The Problem of Modern Flood Control and Water Storage.* New York: Devin-Adair Co., 1954.

Porter, Eliot. *The Place No One Knew: Glen Canyon on the Colorado River.* Edited by David Brower. San Francisco: Sierra Club Books, 1963.

Powell, John Wesley. *The Exploration of the Colorado River.* Abridged from the First Edition of 1875, with an introduction by Wallace Stegner. Chicago: University of Chicago Press, 1957.

Powell, John Wesley. *Report on the Lands of the Arid Region of the United States.* Edited by Wallace Stegner. Cambridge: Belknap Press of Harvard University, 1962.

Reisner, Marc. *Cadillac Desert: The American West and Its Disappearing Water.* New York: Viking Penguin Inc., 1986.

Richardson, Elmo. *Dams, Parks, and Politics: Resource Development and Preservation in the Truman-Eisenhower Era.* Lexington: University of Kentucky Press, 1973.

Stavely, Gaylord. *Broken Waters Sing.* Boston: Little, Brown and Company, 1971.

Stegner, Wallace. (ed.). *This Is Dinosaur: Echo Park Country and Its Magic Rivers.* New York: Alfred Knopf, 1955.

Stegner, Wallace. *Mormon Country.* New York: Duell, Sloan, and Pierce, 1942.

Stegner, Wallace. *Beyond the Hundredth Meridian: John Wesley Powell and the Second Opening of the West.* Boston: Houghton Mifflin, 1954.

Stegner, Wallace. *The Sound of Mountain Water.* New York: Doubleday, 1969.

Stegner, Wallace. *The American West as Living Space.* Ann Arbor: University of Michigan Press, 1987.

Turner, Tom. *Sierra Club: 100 Years of Protecting Nature.* New York: Abrams in association with the Sierra Club, 1991.

Watkins, T.H. *Righteous Pilgrim: The Life and Times of Harold L. Ickes 1874–1952.* New York: Henry Holt and Company, 1990.

Webb, Roy. *Riverman: The Story of Bus Hatch.* Rock Springs, WY: Labyrinth Publishing Co., 1989.

Works Progress Administration. *Utah: A Guide to the State.* New York: Hastings House, 1941. Permission generously granted by the Utah: A Guide to the State Foundation.

Worster, Donald. *Rivers of Empire: Water, Aridity, and the Growth of the American West.* New York: Pantheon, 1985.

Wyant, William K. *Westward in Eden: The Public Lands and the Conservation Movement.* Berkeley: University of California Press, 1982.

Zakin, Susan. *Coyotes and Town Dogs: Earth First! and the Environmental Movement.*

Zaslowsky, Dyan (for the Wilderness Society). *These American Lands: Parks, Politics, and the Public Lands.* New York: Henry Holt, 1986.

Zwinger, Ann. *Run, River, Run.* New York: Harper and Row, 1975.

Acknowledgments

While writing this book I turned to many people for assistance, and I am eager to express my gratitude for the many contributions they have made. The first note of thanks must go to the professors at Brown University who took an interest in this project in its early stages as an honors thesis for the history department. John Thomas diligently directed the development of the thesis over the course of three semesters. The final product owes a great deal to his insight and encouragement. I am also grateful to Caroline Karp of the Department of Environmental Studies at Brown, not only for her valuable comments, but as much for the enthusiasm with which she offered her help.

Park Service employees at Dinosaur National Monument went to great lengths to ensure that my visits there were enjoyable as well as productive. Green River District Naturalist Herm Hoops allowed me to use the historical archives at the monument and made available his own personal file, including a number of documents which were salvaged from the flooded basement of a park service employee. Both collections were invaluable. Mr. Hoops made many other contributions to this project, but I am especially grateful to him for introducing me to the river. Our raft trip through Split Mountain greatly enhanced my appreciation of Dinosaur's canyons and of the impulse to protect them.

Historians do not always have the pleasure of meeting the people about whom they write. I was fortunate to find three people who were directly involved in the Echo Park controversy, and without exception I found their enthusiasm contagious. Don Hatch of Vernal, Utah, wrote extensive letters describing his involvement in Sierra Club raft trips in the 1950s and related interesting anecdotes about his life-long efforts to protect Dinosaur. Charles Eggert provided me with a copy of the film "Wilderness River Trail" and arranged a screening of portions of other films shot in Grand and Glen canyons. In the course of our conversations, Charlie provided much interesting background on the publicity campaign to save Dinosaur. And Dave Brower was exceedingly generous with his time, answering my questions with patience and pointing out some inaccuracies in an early version of the manuscript.

Others who read and commented on early versions of the manuscript include: Richard Millett of the Dinosaur Nature Association, David Whitman and Herm Houps of the National Park Service, and two anonymous reviewers. Their comments were extremely valuable. Walt Borneman's editing enhanced the narrative, and Theresa Duggan ensured that all the final production details came together at the end.

Even though it is not possible to list them all here, I would be remiss not to acknowledge my debt to the scholars whose work provided the foundation for this book. Foremost among these is the late Wallace Stegner. No book on western water management can be written without a deferential bow to Mr. Stegner. Philip Fradkin, Russell Martin, John McPhee, Roderick Nash, Marc Reisner, Elmo Richardson, and Donald Worster have all written books without which my own would have been impossible. I will be pleased if my own work makes some small contribution to the literature from which I have borrowed so liberally. I am especially grateful to those writers and publishers who granted permission to use excerpts from their work.

Finally, no words can express the gratitude I owe my family. To thank them for the support they gave me as I wrote this book would be to acknowledge only the least of my debts to them.

Index